Solid Waste
and Emergency Response
(5102G)

EPA-542-R-05-006
December 2005
http://www.clu-in.org/POPs

Reference Guide to Non-combustion Technologies for Remediation of Persistent Organic Pollutants in Stockpiles and Soil

Internet Address (URL) http://www.epa.gov

CONTENTS

Appendix

Technology Fact Sheets for Remediation of POPs

A   Anaerobic Bioremediation Using Blood Meal for the Treatment of Toxaphene in Soil and Sediment

B   Bioremediation Using DARAMEND® for Treatment of POPs in Soils and Sediments

C   In Situ Thermal Desorption for Treatment of POPs in Soils and Sediments

LIST OF TABLES

ACRONYMS AND ABBREVIATIONS

μg/kg	Microgram per kilogram
ART	Adventus Remediation Technologies, Inc.
BCD	Base-catalyzed decomposition
CCMS	Committee on the Challenges of Modern Society
cy	Cubic yard
DDD	Dichlorodiphenyldichloroethane
DDE	Dichlorodiphenyldichloroethylene
DDT	Dichlorodiphenyltrichloroethane
Dioxins	Polychlorinated dibenzo-p-dioxins
DRE	Destruction and removal efficiency
EDL	Environmental Decontamination Ltd.
EPA	U.S. Environmental Protection Agency
ERT	Environmental Response Team
ESVE	Enhanced soil vapor extraction
FRTR	Federal Remediation Technologies Roundtable
Furans	Polychlorinated dibenzo-p-furans
GEF	Global Environmental Facility
GPCR	Gas-phase chemical reduction
gpd	Gallon per day
GRIC	Gila River Indian Community
HCB	Hexachlorobenzene
HCH	Hexachlorocyclohexane
HEPA	High-efficiency particulate air
ICV	In Container Vitrification
IHPA	International HCH and Pesticides Association
ISTD	In situ thermal desorption
LTR	Liquid tank reactor
M	Meter
MCD	Mechanochemical dehalogenation
MDL	Method detection limit
mg/kg	Milligram per kilogram
mm	Millimeter
ng/kg	Nanogram per kilogram
NAPL	Nonaqueous-phase liquid
NATO	North Atlantic Treaty Organization
NCDENR	North Carolina Department of Environmental and Natural Resources
ND	Below detection limit
OSRTI	Office of Superfund Remediation and Technology Innovation
PACT	Plasma Arc Centrifugal Treatment
PAH	Polycyclic aromatic hydrocarbons
PCB	Polychlorinated biphenyl
PCS	Plasma Converter System
POP	Persistent organic pollutant
ppb	Part per billion
ppm	Part per million
ppt	Part per trillion
REACHIT	Remediation and Characterization Innovative Technologies
SCWO	Supercritical water oxidation
SITE	Superfund Innovative Technology Evaluation

SPHTD	Self-propagating high-temperature dehalogenation
SPV	Subsurface Planar Vitrification
STAP	Science and Technology Advisory Panel
SVOC	Semivolatile organic compound
TCDD	Tetrachlorodibenzodioxin
TNT	Trinitrotoluene
UNEP	United Nations Environment Programme
UNR	University of Nevada at Reno
USD	United States Dollar
VOC	Volatile organic compound

NOTICE AND DISCLAIMER

This report compiles information about non- combustion technologies for remediation of persistent organic pollutants, including technology applications at both domestic and international sites, but is not a comprehensive review of all the current non- combustion technologies or vendors. This report also does not provide guidance regarding selection of a specific technology or vendor. Use or mention of trade names or commercial products does not constitute endorsement or recommendation for use.

This report has undergone EPA and external review by experts in the field. However, information in this report is derived from a variety of references (including personal communications with experts in the field), some of which have not been peer-reviewed.

This report has been prepared by the U.S. Environmental Protection Agency (EPA) Office of Superfund Remediation and Technology Innovation, with support provided under Contract Number 68-W-02-034. For further information about this report, please contact Ellen Rubin at EPA's Office of Superfund Remediation and Technology Innovation, at (703) 603-0141, or by e-mail at rubin.ellen@epa.gov.

A PDF version of "Non- combustion Technologies for Remediation of Persistent Organic Pollutants in Stockpiles and Soil" is available for viewing or downloading at the Hazardous Waste Cleanup Information System web site at http://www.clu-in.org/POPs. A limited number of printed copies of the report are available free of charge and may be ordered via the web site, by mail, or by fax from the following source:

EPA/National Service Center for Environmental Publications
P.O. Box 42419
Cincinnati, OH 45242-2419
Telephone: (513) 489-8190 or (800) 490-9198
Fax: (513) 489-8695

ACKNOWLEDGMENTS

Special acknowledgment is given to members of the International HCH and Pesticides Association (IHPA), and other remediation professionals for their cooperation, thoughtful suggestions, and support during the preparation of this report. Contributors to the report include the following individuals:

Bryan Black, Environmental Decontamination Ltd.
Carl V. Mackey, Washington Group International
Charles Rogers, BCD Group, Inc.
Christine Parent, California Department of Toxic Substances Control
David Raymond, Adventus Remediation Technologies, Inc.
Edward Someus, Terra Humana Clean Technology Engineering Ltd.
Giacomo Cao, Centro Studi Sulle Reazioni Autopropaganti
John Vijgen, International HCH and Pesticides Association
Kevin Finucane, AMEC Earth and Environmental, Inc.
Matt Van Steenwyk, CerOx™
Paul Austin, Sonic Environmental Solutions, Inc.
Ralph Baker, TerraTherm, Inc.
Tedd E. Yargeau, California Department of Toxic Substances Control
Volker Birke, Tribochem

EXECUTIVE SUMMARY

This report provides a high level summary of information on the applicability of existing and emerging non-combustion technologies for the remediation of persistent organic pollutants (POPs) in stockpiles and soil. POPs are a set of chemicals that are toxic, persist in the environment for long periods of time, and biomagnify as they move up through the food chain. POPs have been linked to adverse effects on human health and animals, such as: cancer, damage to the nervous system, reproductive disorders, and disruption of the immune system. In addition, restrictions and bans on the use of POPs have resulted in a significant number of unusable stockpiles of POP-containing materials internationally. Deterioration of storage facilities used for the stockpiles, improper storage practices, and past production and use of POPs also have resulted in contamination of soils around the world.

Previously, POPs have been destroyed by combustion technologies (incineration). Many interested parties have expressed concern about the potential environmental and health effects associated with this type of treatment technology. Combustion of POPs can create by-products such as polychlorinated dibenzo-p-dioxins (dioxins) and polychlorinated dibenzo-p-furans (furans) - known human carcinogens. Two principal reports have identified the various non-combustion destruction technologies for POPs (*Review of Emerging, Innovative Technologies for the Destruction and Decontamination of POPs and the Identification of Promising Technologies for Use in Developing Countries; Evaluation of Demonstrated and Emerging Remedial Action Technologies for the Treatment of Contaminated Land and Groundwater (Phase III)*).

With the passage of time, some of the technologies discussed in these comprehensive documents that were in the development stage are now commercialized; while other commercial technologies are no longer being developed. Also, new promising destruction technologies for POPs have been developed. This report is intended to summarize and update older reports in a concise reader's guide, with links to sources of further information. The updated information was obtained by reviewing various websites and documents, and by contacting technology vendors and experts in the field.

This report provides short descriptions of a range of non-combustion technologies and highlights new performance data showing the various considerations associated with selecting a non-combustion technology. Table 3-1 summarizes the selected technologies and provides information on waste strength treated, ex situ or in situ technology, contaminants treated, cost information when available, pretreatment requirements, power requirements, configuration needs, and links to individual fact sheets. Fact sheets for the various technologies are available through the International Hexachlorocyclohexane (HCH) and Pesticides Association website; new fact sheets are available in the appendices of this report.

.

1.0   INTRODUCTION

POPs are toxic compounds that are chemically stable, do not easily degrade in the environment, and tend to accumulate and biomagnify as they move up through the food chain. Serious human health problems are associated with POPs, including cancer, neurological damage, birth defects, sterility, and immune system suppression. Restrictions and bans on the use of POPs have resulted in a significant number of unusable stockpiles of POP-containing materials internationally. In addition, deterioration of storage facilities used for the stockpiles, improper storage practices, and past production and use of POPs have resulted in contamination of soils around the world. Because of their chemical stability, tendency to bioaccumulate, adverse health effects associated with POPs, and widespread POP contamination, remediation technologies are needed to treat these pollutants.

Previously, POP-contaminated soil and stockpiles have been treated using technologies such as incineration that rely on combustion to destroy the contaminants. However, site owners and operators, remedial project managers, and other interested parties have expressed concern about the potential environmental and health effects associated with combustion of POPs. Combustion technologies can create polychlorinated dibenzo-p-dioxins (dioxins) and polychlorinated dibenzo-p-furans (furans). Dioxins and furans have been characterized by EPA as human carcinogens and are associated with serious human health problems. Also, combustion technologies that have historically been used for the destruction of POPs may fail to meet the stringent environmental conditions or destruction and removal efficiency (DRE) requirements established for POPs. Because of these concerns and an ongoing desire to find more cost effective solutions, environmental professionals are examining the application of non-combustion technologies to remediate POPs in stockpiles and soil (Ref. 58).

Under the Stockholm Convention, countries committed to reduce or eliminate the production, use, and release of the 12 POPs of greatest concern to the global community. In addition, the Basel Convention invited the bodies of the Stockholm Convention to consider the development of information on best available techniques and environmental practices with respect to POPs (Refs. 59 and 60). The Basel Convention was adopted on March 22, 1989, by the Conference of Plenipotentiaries convened at Basel. The Stockholm Convention obligated parties to remediate POPs stockpiles but did not obligate cleanup of POPs-contaminated sites. Table 1-1 lists the 12 specific POPs identified by the Stockholm Convention, which include nine pesticides and three industrial chemicals or by-products (Ref. 23).

Table 1-1. POPs Identified by the Stockholm Convention

Pesticides	Industrial Chemicals or By-Products
Aldrin	Polychlorinated biphenyls (PCB)
Chlordane	Dioxins
Dichlorodiphenyltrichloroethane (DDT)	Furans
Dieldrin	
Endrin	
Heptachlor	
Hexachlorobenzene (HCB)	
Mirex	
Toxaphene	

Source: Ref. 23

1.1 Purpose of Report

This report is intended to provide a high level summary of information for federal, state, and local regulators, site owners and operators, consultants, and other stakeholders on the applicability of existing and emerging non-combustion technologies for the remediation of POPs in stockpiles and soil. The report provides short descriptions of these technologies and evaluates them based on the POPs treated, media treated, pretreatment requirements, performance and cost. Case studies provided show the various considerations associated with selecting a non- combustion technology.

Information on non- combustion technologies for the remediation of POPs is available in several more comprehensive documents. With the passage of time, some of the technologies discussed in these comprehensive documents were in the development stage and are now commercialized; while other commercial technologies are no longer being developed. For all of these technologies, this report is intended to update and summarize older reports in a relevant concise reader's guide with links to sources of further information. In addition, this report provides information on several new technologies.

1.2 Methodology

INTERNATIONAL HCH AND PESTICIDES ASSOCIATION
IN 2002, JOHN VIJGEN, THROUGH THE INTERNATIONAL HCH AND PESTICIDES ASSOCIATION, PUBLISHED 11 FACT SHEETS ABOUT EMERGING NON- COMBUSTION ALTERNATIVES FOR THE ECONOMICAL DESTRUCTION OF POPS (HTTP://WWW.IHPA.INFO/LIBRARYNATO.HTM). THESE FACT SHEETS WERE USED AS A KEY INFORMATION SOURCE DURING DEVELOPMENT OF THIS REPORT.

EPA identified non- combustion technologies for remediation of POPs in stockpiles and soil by reviewing technical literature, EPA reports, and EPA databases such as the Federal Remediation Technologies Roundtable (FRTR) (www.frtr.gov) and the Remediation and Characterization Innovative Technologies (REACHIT) system (www.epareachit.org), as well as by contacting technology vendors and experts in the field. REACHIT is a real time vendor supplied source of information including data on emerging non-combustion technologies for POPs. A key source of information is the work done by John Vijgen of the International HCH and Pesticides Association (see box). Some of the information sources have not been peer-reviewed.

A list of non- combustion technologies was prepared using the available information. For each technology, the following types of information were compiled: commercial availability; the processes used; advantages and limitations; POPs treated; sites where the technology was applied at full-, pilot- or bench-scale; technology performance results; cost information; and lessons learned. Technologies that have treated one or more of the 12 POPs or have the potential to treat POPs are discussed in this report. Some technologies previously discussed in other sources were identified as no longer commercially available or have not been used to treat POPs.

Based on the available information, EPA reviewed the types of waste and contaminants treated, and summarized the results from use of the technology. Performance data were evaluated based on the concentrations of specific POPs before-and after-treatment. For many of the specific projects described in this report, gaps existed in the information available. For example, for some projects, little or no performance data was available. EPA did not perform independent evaluations of technology performance. However, where feasible, such data gaps were addressed by contacting specific technology vendors and users.

1.3   Report Organization

This report includes six sections. Section 1.0 is an introduction discussing the purpose and organization of the report. Section 2.0 provides background information about the Stockholm Convention and about the sources, characteristics, and health effects of POPs. Section 3.0 presents technology overviews, while more detailed information for some technologies is provided in the technology-specific fact sheets in the appendices of the report. Seventeen technologies for POP treatment are described in Section 3.0. Section 3.0 is divided into four

> **FURTHER INFORMATION ABOUT NON-COMBUSTION TECHNOLOGIES FOR REMEDIATION OF POPs IS PROVIDED AT WWW.CLU-IN.ORG/POPS.**

subsections based on the scale of application of the technologies. Section 3.1 contains descriptions of full-scale technologies that treat POPs. Section 3.2 contains descriptions of pilot-scale technologies that have treated POPs. Section 3.3 contains descriptions of bench-scale technologies that have been tested on POPs. Section 3.4 contains descriptions of full-scale technologies that have treated non-POPs and that are potentially applicable for POP treatment. Section 4.0 lists web-based information sources used for the preparation of this report. Section 5.0 contains contact details for technology vendors. Section 6.0 lists references used in the preparation of this report.

The appendices to this report provide fact sheets prepared by EPA for three technologies: anaerobic bioremediation using blood meal for the treatment of toxaphene in soil, DARAMEND® technology for treatment of POPs in soils, and in situ thermal desorption (ISTD) for treatment of POPs in soil. Fact sheets for 11 other POP treatment technologies discussed in this report were previously published in "Evaluation of Demonstrated and Emerging Remedial Action Technologies for the Treatment of Contaminated Land and Groundwater (Phase III)," which was issued by the International HCH and Pesticides Association (IHPA) in 2002. EPA examined the 11 technologies for which fact sheets were prepared by IHPA (see list in Section 2.5) and evaluated whether additional, more recent information was available for these technologies. Only one technology, mechanochemical dehalogenation (MCD), was identified for which new information had become available after the original fact sheet was published; this new information is included in Section 3.1.6 of this report. All other full-scale technologies listed in this report were updated with site specific performance data and included in their respective sections.

2.0   BACKGROUND

This section provides background information about the Stockholm Convention and the sources, characteristics, and health effects of POPs. It also identifies related documents that address technologies for the treatment of POPs.

2.1   Stockholm Convention

The Stockholm Convention (Ref. 49) is a global treaty intended to protect human health and the environment from POPs. On May 23, 2001, 93 countries and regional economic integration organizations such as the European Union signed the convention. As of April 25, 2005, 97 countries and one regional economic integration organization had signed or ratified the treaty. The United States signed the treaty but as of April 2005 has not ratified it.

2.2   Sources of POPs

Most POPs originate from man-made sources associated with production, use, and disposal of certain organic chemicals. Some POPs are intentionally produced, while others are the by-products of industrial processes or result from the combustion of organic chemicals. The POPs within the scope of the Stockholm Convention include nine pesticides and three industrial chemicals or by-products (Ref. 18). Table 1-1 lists these POPs.

The nine pesticides targeted by the Stockholm Convention were produced intentionally and used on agricultural crops or for public health vector control. By the late 1970s, these pesticides had been either banned or subjected to severe use restrictions in many countries. However, some of the pesticides are still in use in parts of the world where they are considered essential for protecting public health (Ref. 18).

The three industrial chemicals and by-products within the scope of the Stockholm Convention are PCBs, dioxins, and furans. PCBs were produced intentionally but are typically released into the environment unintentionally. The most significant use of PCBs was as a dielectric fluid (a fluid which can sustain a steady electrical field and act as an electrical insulator) in transformers and other electrical and hydraulic equipment. Most countries stopped producing PCBs in the 1980s; for example, equipment manufactured in the United States after 1979 usually does not contain PCBs. However, older equipment containing PCBs is still in use. Most capacitors manufactured in the United States before 1979 also contain PCBs.

Dioxins and furans are usually produced and released unintentionally. They may be generated by industrial processes or by combustion, including fuel burning in vehicles, municipal and medical waste incineration, open burning of trash, and forest fires (Ref. 18).

2.3   Characteristics of POPs

POPs are synthetic chemicals with the following properties (Ref. 18):

- They are toxic and can have adverse effects on human health and animals.
- They are chemically stable and do not readily degrade in the environment.
- They are lipophillic (affinity for fats) and easily soluble in fat.
- They accumulate and biomagnify as they move up through the food chain.
- They move over long distances in nature and can be found in regions far from their points of origin or use.

2.4   Health Effects of POPs

POPs are associated with serious human health problems, including cancer, neurological damage, birth defects, sterility, and immune system defects. EPA has classified certain POPs as probable human carcinogens[1], including aldrin, dieldrin, chlordane, DDT, heptachlor, HCB, toxaphene, and PCBs. Laboratory studies have shown that low doses of POPs can adversely affect organ systems. Chronic exposure to low doses of certain POPs may affect the immune and reproductive systems. Exposure to high levels of certain POPs can cause serious health effects or death. The primary potential human health effects associated with POPs are listed below (Refs. 18 and 56).

- Cancer
- Immune system suppression
- Nervous system disorders
- Reproductive damage
- Altered sex ratio
- Reduced fertility
- Birth defects
- Liver, thyroid, kidney, blood, and immune system damage
- Endocrine disruption
- Developmental disorders
- Shortened lactation in nursing women
- Chloracne and other skin disorders

In addition, studies have linked POP exposure to diseases and abnormalities in a number of wildlife species, including numerous species of fish, birds, and mammals. For example, in certain birds of prey, high levels of DDT caused eggshells to thin to the point that the eggs could not produce live offspring (Ref. 18).

2.5   Related Documents

Two organizations, the United Nations Environment Programme (UNEP) and IHPA, have recently developed summary overview reports and fact sheets about non- combustion technologies for POP treatment. These documents are listed below.

- UNEP, Science and Technology Advisory Panel (STAP) of the Global Environmental Facility (GEF). 2004. "Review of Emerging, Innovative Technologies for the Destruction and Decontamination of POPs and the Identification of Promising Technologies for Use in Developing Countries." GF/8000-02-02-2205. January. Online Address: http://www.basel.int/techmatters/review_pop_feb04.pdf. This report (Ref. 57) provides a summary overview of non- combustion technologies that are considered to be innovative and emerging and that have been identified as potentially promising for the destruction of POPs in stockpiles. The report was originally a background document for the STAP-GEF workshop held in Washington, DC, in October 2003 and was based on work done by the International Centre for

---

[1] Based on the 1986 EPA classification of carcinogens, "probable" carcinogens (Group B) include those agents for which the weight of evidence of human carcinogenicity based on epidemiological studies is "limited" and those agents for which the weight of evidence of human carcinogenicity based on animal studies is "sufficient" (Ref. 56).

Sustainability Engineering and Science, Faculty of Engineering, at the University of Auckland, New Zealand.

The report contains overviews of the following non-combustion technologies:

1. Base-catalyzed decomposition (BCD)
2. Bioremediation/Fenton reaction
3. Catalytic hydrogenation
4. DARAMEND® bioremediation
5. Enzyme degradation
6. Fe (III) photocatalyst degradation
7. Gas-phase chemical reduction (GPCR)
8. GeoMelt™ process
9. In situ bioremediation of soils
10. Mechanochemical dehalogenation (MCD)
11. Mediated electrochemical oxidation (AEA Silver II)
12. Mediated electrochemical oxidation (CerOx™)
13. $MnO_x/TiO_2 – Al_2O_3$ catalyst degradation
14. Molten metal
15. Molten salt oxidation
16. Molten slag process
17. Ozonation/electrical discharge destruction
18. Photochemically enhanced microbial degradation
19. Phytoremediation
20. Plasma arc (PLASCON™)
21. Pyrolysis
22. Self-propagating high-temperature dehalogenation (SPHTD)
23. Sodium reduction
24. Solvated electron technology
25. Supercritical water oxidation (SCWO)
26. TiO_2 – based V_2O_5/WO_3 catalysis
27. White rot fungi bioremediation

- IHPA. 2002. IHPA and North Atlantic Treaty Organization (NATO) Committee on the Challenges of Modern Society (CCMS) Pilot Study Fellowship Report: "Evaluation of Demonstrated and Emerging Remedial Action Technologies for the Treatment of Contaminated Land and Groundwater (Phase III)." Online Address: http://www.ihpa.info/libraryNATO.htm . This report (Ref. 33) describes emerging non- combustion alternatives for the economical destruction of POPs. Mr. John Vijgen of IHPA collected the technology data and authored the report. The report contains fact sheets for the 11 technologies listed below:

1. BCD
2. CerOx™
3. Gas-phase chemical reduction process
4. GeoMelt™
5. In situ thermal destruction
6. MCD™
7. SPHTD
8. Silver II™
9. Solvated electron technology
10. SCWO
11. TDR-3R™

3.0   NON-COMBUSTION TECHNOLOGIES

This section provides a review of selected non- combustion technologies for POPs remediation, including their implementation at both domestic and international sites. In this report, POPs include the 12 contaminants within the scope of the Stockholm Convention, and non-combustion technologies are defined as processes that operate in a starved or ambient oxygen atmosphere (including thermal processes). For this report, treatment technology is defined as the primary process where contaminant destruction occurs. Pretreatment is defined as any process that precedes the primary treatment technology wherein the contaminants are transferred from one media/phase to another.

Table 3-1 lists the selected technologies and summarizes available technology-specific information, including capability to handle waste strength, ex situ or in situ application, scale, contaminant treated, cost, pre-treatment needs, power requirements, configuration and location of fact sheets. Waste strength refers to high- and low-strength wastes. High-strength waste includes stockpiles of POPs-contaminated materials and highly contaminated soil. Low-strength waste includes soil contaminated with low concentrations of POPs. The table indicates whether the technologies have been applied at a full[2], pilot[3], or bench[4] scale for treatment of POPs. Table 3-2 provides performance data for the selected technologies. The performance data include the site location, contaminants treated, untreated and treated contaminant concentration, and percent reduction of the contaminants. Section 5.0 provides contact information for vendors of these various technologies.

3.1   Full-Scale Technologies for Treatment of POPs

This section describes seven technologies that have been implemented to treat POPs at full scale. Each subsection focuses on a single technology and includes a description of the technology and information about its application at specific sites. Fact sheets developed by EPA and IHPA contain additional details on some of these technologies and their applications. The appendices to this report provide fact sheets prepared by EPA for three technologies. Links to the IHPA fact sheets are included in the appropriate subsections of this report.

---

[2] A full-scale project involves use of a commercially available technology to treat industrial waste and to remediate an entire area of contamination.
[3] A pilot-scale project is usually conducted in the field to test the effectiveness of a technology and to obtain information for scaling up a treatment system to full scale.
[4] A bench-scale project is conducted on a small scale, usually in the laboratory, to evaluate a technology's ability to treat soil, waste, or water. Such a project often occurs during the early phases of technology development.

Reference Guide to Non-combustion Technologies for Remediation of Persistent Organic Pollutants in Stockpiles and Soil

Table 3-1. Summary of Non-combustion Technologies for Remediation of Persistent Organic Pollutants [1]

| Technology | Waste Strength [2] | Ex/In situ [3] | Contaminant(s) Treated | | | Non-POPs [5] | Cost | Pre-Treatment | Power Requirement | Configuration | Fact Sheet |
| | | | POPs | | | | | | | | |
			Pesticide(s) [4]	PCBs	Dioxin/Furans						
Full-Scale Technologies											
Anaerobic bioremediation using blood meal for the treatment of toxaphene in soil and sediment	Low	Ex situ	Toxaphene	None	None	None	$98 to $296 per cubic yard (in 2004)	None	None	Transportable	Appendix A
DARAMEND®	Low	Ex/In situ	Toxaphene and DDT	None	None	DDD, DDE, RDX, HMX, DNT, and TNT	$55 per cubic yard (in 2004)	None	None	Transportable	Appendix B
Gas Phase Chemical Reduction (GPCR™) [6]	High	Ex situ	DDT and HCB	Yes	Yes	PAH, chlorobenzene	NA	Thermal desorption	High	Fixed and transportable	http://www.ihpa.info/library nato.htm
GeoMelt™	Low/High	In/Ex situ	DDT, chlordane, dieldrin and HCB	Yes	Yes	Metals and radioactive waste	NA	None	High	Fixed and transportable	http://www.ihpa.info/library nato.htm
In-Situ Thermal Desorption (ISTD)	Low/High	In situ	NA	Yes	Yes	VOCs, SVOCs, oils, creosote, coal tar, gasoline, MTBE, volatile metals	$200 to $600 per cubic yard (from 1996 to 2005)	None	High	Transportable	Appendix C
Mechanochemical Dehalogenation (MCD™)	High	Ex situ	Aldrin, dieldrin and DDT	None	None	Lindane, DDD, and DDE	NA	None	NA	NA	http://www.ihpa.info/library nato.htm
Xenorem™	Low	Ex situ	Chlordane, DDT, dieldrin, and toxaphene	None	None	Molinate	$132 per cubic yard (in 2000)	None	NA	Transportable	None

Reference Guide to Non-combustion Technologies for Remediation of Persistent Organic Pollutants in Stockpiles and Soil

Technology	Waste Strength [2]	Ex/In situ [3]	Contaminant(s) Treated				Cost	Pre-Treatment	Power Requirement	Configuration	Fact Sheet
			POPs			Non-POPs [5]					
			Pesticide(s) [4]	PCBs	Dioxin/Furans						
Pilot-Scale Technologies											
Base Catalyzed Decomposition (BCD)	Low/High	Ex situ	Chlordane and heptachlor	Yes	Yes	a-BHC, endosulfan	NA	Thermal desorption	High	Transportable and fixed	http://www.ihpa.info/library.nato.htm
CerOx™	Low	Ex situ	Chlordane	Yes	Yes	Aniline, cyclohexanone, and dibutyl phthalates	NA	Blending to produce liquid influent	NA	Modular	http://www.ihpa.info/library.nato.htm
Phytoremediation	Low	In/Ex situ	DDE, DDT, and chlordane	Yes	None	NA	NA	None	None	Transportable	None
Sonic Technology	Low/High	Ex situ	None	Yes	None	NA	NA	Mixing with solvent to produce a slurry	NA	NA	None
Bench-Scale Technologies											
Self Propagating High Temperature Dehalogenation	High	Ex situ	HCB	None	None	None	NA	None	NA	NA	http://www.ihpa.info/library.nato.htm
TDR-3R™	High	Ex situ	HCB	None	None	PAH	NA	Thermal desorption	High	NA	http://www.ihpa.info/library.nato.htm

Notes:

1: Data in this table is derived from various document, vendor information, and other sources - both peer reviewed and not.

2: Waste strength refers to high- and low-strength wastes. High-strength waste includes stockpiles of POP-contaminated materials and highly contaminated soil. Low-strength waste includes soil contaminated with low concentrations of POPs.

3: Ex/In situ refers to Ex situ or In situ application of the technology.

4: Pesticides include the nine pesticides addressed within the scope of the Stockholm Convention.

5: Non-POPs include contaminants outside the scope of Stockholm Convention.

6: GPCR is currently not commercialized due to cost.

BHC:	Benzene hexachloride
DDD:	Dichlorodiphenyldichloroethane
DDE:	dichlorodiphenyldichloroethylene
DDT:	dichlorodiphenyltrichloroethane
DNT:	Di-nitro toluene
HMX:	High melting explosive, octahydro-1,3,5,7-tetranitro-1,3,5,7 tetrazocine

MTBE:	Methyl tert-butyl ether
NA:	Not available
PAH:	Polycyclic aromatic hydrocarbons
SVOC	Semivolatile organic compound
VOC	Volatile organic compound

Table 3-2. Performance of Non-combustion Technologies for Remediation of Persistent Organic Pollutants [1]

Technology	Site Name or Location	Contaminant	Untreated Concentration (mg/kg)	Treated Concentration (mg/kg)	Percent Reduction
		Examples of Treatment Performance[2]			
Full-Scale Technologies					
Anaerobic bioremediation using blood meal for the treatment of toxaphene in soil and sediment	Gila River Indian Community, Arizona	Toxaphene	59	4	93%
DARAMEND®	T.H. Agricultural and Nutrition Superfund Site, Montgomery, Alabama	Toxaphene	189	21	89%
		DDT	84.5	8.65	90%
Gas Phase Chemical Reduction (GPCR™)[3]	NA	NA	NA	NA	NA
GeoMelt™	Parsons Chemical Superfund Site, Grand Ledge, Michigan	DDT	340	< 4	99%
		Chlordane	89	< 1	99%
		Dieldrin	4.6	< .008	99%
In-Situ Thermal Desorption (ISTD)	Tanapag Village, Saipan, Northern Mariana Islands	PCBs	10,000 (max)	< 1	99.99%
	Centerville Beach, Ferndale, California	PCBs	860 (max)	< 0.17	99.98%
		Dioxin/Furans	0.0032	0.00006	99.81%
Mechanochemical Dehalogenation (MCD™)	Fruitgrowers Chemical Company Site, Mapua, New Zealand	Aldrin	7.52	0.798	91%
		Dieldrin	65.6	19.8	70%
		DDX (total DDT, DDD, and DDE)	717	64.8	91%
Xenorem™	Stauffer Management Company Superfund Site, Tampa, Florida	Chlordane	3.8	< MDL[4]	NA
		DDT	82	9.8	88%
		Dieldrin	2.4	< MDL[4]	NA
		Toxaphene	129	7.8	94%
Pilot-Scale Technologies					
Base Catalyzed Decomposition (BCD)	Warren County Landfill, Warren County, North Carolina	PCBs	81,100	< 5	99.99%
	FCX Superfund Site, Statesville, North Carolina	Heptachlor	0.648	ND[2]	NA
		Chlordane	9.7173	ND[2]	NA
CerOx™	NA	NA	NA	NA	NA
Sonic Technology	Juker Holdings Site, Vancouver, British Columbia	PCBs	NA	NA	NA
Bench-Scale Technologies					
Self Propagating High Temperature Dehalogenation	NA	NA	NA	NA	NA
TDR-3R™	Gare Site, Hungary	HCB	1,215	0.1	99.99%

Notes:

1: Data in this table is derived from various document, vendor information, and other sources - both peer reviewed and not.
2: Treatment examples were selected to illustrate the types of treatment performance data available.
3: GPCR is currently not commercialized due to cost.
4: The specific limits for the MDL and ND were not provided in the source document.

DDD:	Dichlorodiphenyldichloroethane	PCBs:	Polychlorinated biphenyls	NA:	Not available
DDE:	Dichlorodiphenyldichloroethylene	ND:	Below detection limit		
DDT:	Dichlorodiphenyltrichloroethane	MDL:	Method detection limit		
HCB:	Hexachlorobenzene	mg/kg:	Milligram per kilogram		

3.1.1 Anaerobic Bioremediation Using Blood Meal for Treatment of Toxaphene in Soil and Sediment

This technology uses biostimulation with amendments to promote degradation of toxaphene in soil or sediment by native anaerobic microorganisms. It involves the addition of biological amendments such as blood meal (dried and powdered animal blood), which is used as a nutrient, and phosphates, which are used as a pH buffer (Ref. 2). In some applications, starch is also used to speed the establishment of anaerobic conditions. The soil to be treated is mixed with the amendments and water. The technology can use several methods to produce homogeneous soil-amendment mixtures, including blending in a dump truck, mechanical mixing in a pit, and mixing in a pug mill. The homogenized mixture is transferred to a lined cell, and water is added to produce a slurry. Up to a foot of water cover is provided above the settled solids. The water cover helps to minimize the transfer of atmospheric oxygen to the slurry so that anaerobic conditions are maintained. The lined cell is covered with a plastic sheet, and the slurry is incubated for several months.

The slurry may be sampled periodically to measure contaminant degradation. The process continues until the treatment goals are achieved, at which time the cell is drained. The treated slurry is usually left in the cell; however, the slurry may be dried and used as fill material on site or as a source of microorganisms for other applications of the technology. The end products of degradation are carbon dioxide, water, and chlorides. Residual contamination can include low concentrations of toxaphene and camphenes with varying degrees of chlorination (Refs. 3 and 19).

Anaerobic bioremediation using blood meal has been implemented to treat low-strength waste contaminated with toxaphene. Essential components such as mixing troughs are typically constructed and left in place. Other components such as mixing equipment and biological amendments are usually procured locally.

> **THE FACT SHEET PREPARED BY EPA IS INCLUDED IN APPENDIX A.**

POPs TREATED: TOXAPHENE

MEDIUM: SOIL AND SEDIMENT

RESIDUALS: LOW CONCENTRATIONS OF TOXAPHENE AND CAMPHENES WITH VARYING DEGREES OF CHLORINATION

COSTS: $98 TO $296 PER CUBIC YARD (COST IN 2004 USD)

*FULL SCALE
EX SITU*

The technology has been used to treat toxaphene at numerous livestock dip vat sites. Dip vats are trenches with a pesticide formulation used to treat livestock infested with ticks. In 2004, cleanup costs in United States Dollar (USD) for full-scale implementations ranged from $98 to $296 per cubic yard (Ref. 19). Performance data from nine dip vat site applications are presented in Table 3-3.

The technology was developed by EPA's Environmental Response Team (ERT). This technology is publicly available and is not patented (Ref. 3).

Table 3-3. Performance of Anaerobic Bioremediation Using Blood Meal for Toxaphene Treatment

Site	Location	Period (Days)	Quantity of Soil Treated	Scale	Untreated Concentration (mg/kg)	Treated Concentration (mg/kg)
Gila River Indian Community (GRIC)						
GRIC Cell 1	Chandler, Arizona	272	3,500 cy	Full	59	4
GRIC Cell 2		272		Full	31	4
GRIC Cell 3		272		Full	29	2
GRIC Cell 4		272		Full	211	3
Navajo Vats Chapter						
Laahty Family Dip Vat	Zuni Nation, New Mexico	31	253 cy	Full	29	4
Henry O Dip Vat	Zuni Nation, New Mexico	68	660 cy	Full	23	8
Nazlini	NA	108	3.5 tons	Pilot	291	71
Whippoorwill	NA	110	3.5 tons	Pilot	40	17
Blue Canyon Road	NA	106	NA	NA	100	17
Jeddito Island	NA	76	NA	NA	22	3
Ojo Caliente	Zuni Nation, New Mexico	14	200 cy	NA	14	4
Poverty Tank	NA	345	NA	NA	33	8

Sources: Refs. 2, 3 and 19

Notes:
cy = Cubic yard
mg/kg = Milligram per kilogram
NA = Not available

3.1.2   DARAMEND®

DARAMEND® has been used to treat low-strength wastes contaminated with toxaphene and DDT. The DARAMEND® technology can be implemented ex situ or in situ. It is an amendment-enhanced bioremediation technology for POP treatment that involves the creation of sequential anoxic and oxic conditions (Ref. 41). The treatment process involves the following steps:

1. Addition of a solid-phase DARAMEND® organic soil amendment of a specific particle size distribution and nutrient profile, zero valent iron, and water to produce anoxic conditions
2. Periodic tilling of the soil to promote oxic conditions
3. Repetition of the anoxic-oxic cycle until cleanup goals are achieved

Bioremediation using DARAMEND® process. Source: Ref. 1

The addition of the DARAMEND® organic amendment, zero valent iron, and water stimulates the biological depletion of oxygen, generating strong reducing (anoxic) conditions in the soil matrix. Diffusion of replacement oxygen into the soil matrix is prevented by near saturation of the soil pores with water. The depletion of oxygen creates a very low redox potential, which promotes dechlorination of organochlorine compounds. The soil matrix consisting of contaminated soil and the amendments is left undisturbed for the duration of the anoxic phase of the treatment cycle (typically 1 to 2 weeks). In the next (oxic) phase, periodic tilling of the soil increases diffusion of oxygen and distribution of irrigation water in the soil. The dechlorination products formed during the anoxic degradation process are subsequently removed through aerobic (oxic) biodegradation processes, which are initiated and promoted by the passive air drying and tilling of the soil. Addition of the DARAMEND® amendment and the anoxic-oxic cycle continue until cleanup goals are achieved (Ref. 15).

The DARAMEND® technology can be implemented ex situ or in situ. In both cases, the treatment layer is 2 feet (ft) deep, which is the typical depth reached by tilling equipment. For treatment to greater depths, the technology can be

> **THE FACT SHEET PREPARED BY EPA IS INCLUDED IN APPENDIX B.**

implemented in sequential, 2-ft lifts. The DARAMEND® technology may be technically or economically infeasible with excessively high contaminant concentrations in soils (Ref. 15).

DARAMEND® has been used to treat soil and sediment containing low concentrations of pesticides such as toxaphene and DDT as well as other contaminants. The technology has not been used for treatment of other POPs such as PCBs, dioxins, or furans. Adventus Remediation Technologies, Inc. (ART), the developer of the technology, indicated that DARAMEND® had not been successful in bench-scale treatment of PCB-contaminated soil. DARAMEND® has been used to treat POPs at the T.H. Agriculture and Nutrition Superfund site in Montgomery, Alabama, and the W.R. Grace

> **POPs TREATED: TOXAPHENE AND DDT**
>
> **MEDIUM: SOIL AND SLURRY**
>
> **COSTS: $55 PER TON (COST IN 2004 USD)**
>
> *FULL SCALE*
> *EX SITU AND IN SITU*

site in Charleston, South Carolina. Table 3-4 presents the performance data from these applications. The average treatment cost (in 2004 USD) at the site in Montgomery was $55 per ton; the vendor did not specify the components included in this cost (Refs. 1, 22 and 42).

Table 3-4. Performance of DARAMEND® Technology

Site	Location	Year Implemented	Period (Months)	POP	Quantity of Soil Treated (Tons)	Scale	Untreated Concentration (mg/kg)	Treated Concentration (mg/kg)
T.H. Agriculture and Nutrition Superfund site	Montgomery, Alabama	2003	5	Toxaphene	4,500	Full	189	21
				DDT			84.5	8.65
W.R. Grace site	Charleston, South Carolina	1995	8	Toxaphene	250	Pilot	239	5.1
				DDT			89.7	16.5

Source: Ref. 1

Notes:
mg/kg = Milligram per kilogram

DARAMEND® is a proprietary technology provided by ART in Mississauga, Ontario, Canada. In the United States, the technology is provided by ART's sister company, Adventus Americas, Inc. in Bloomingdale, Illinois.

3.1.3   Gas-Phase Chemical Reduction

Gas-phase chemical reduction (GPCR™) has been used to treat high-strength wastes containing POPs. GPCR™ is an ex situ technology and is available in both fixed and transportable configurations. It is applicable to both solids and liquids.

POPs TREATED: HCB, DDT, PCBs, DIOXINS, AND FURANS

PRETREATMENT: THERMAL DESORPTION

MEDIUM: SOIL, SEDIMENT, AND LIQUID WASTE

FULL SCALE
EX SITU

The technology uses a two-stage process to treat soil contaminated with POPs. In the first stage, contaminated soil is heated in a thermal reduction batch processor in the absence of oxygen to temperatures around 600 °C. This causes organic compounds to desorb from the solid matrix and enter the gas phase. The treated soil is non- hazardous and is allowed to cool prior to its disposal on or off site. In the second stage, the desorbed gaseous-phase contaminants pass to a GPCR™ reactor, where they react with introduced hydrogen gas at temperatures ranging from 850 to 900 °C. This reaction converts organic contaminants into primarily methane and water. Acid gases such as hydrogen chloride may also be produced when chlorinated organic contaminants are present. The gases produced in the second stage are scrubbed by caustic scrubber towers to cool the gases, neutralize acids, and remove fine particulates. The off-gas exiting the scrubber is rich in methane and is collected and stored for reuse as fuel. Methane is also used to generate hydrogen for the GPCR™ process in a catalyzed high-temperature reaction. Spent scrubber water is treated by granular activated carbon filters prior to its discharge (Refs. 12 and 33).

THE FACT SHEET PREPARED BY IHPA IS AVAILABLE AT HTTP://WWW.IHPA.INFO/LIBRARYNATO.HTM.

GPCR™ has been implemented at both full and pilot scales to treat solids and liquids contaminated with POPs. The POPs treated include HCB, DDT, PCBs, dioxins, and furans. Table 3-5 presents performance information for the technology. In 1992, GPCR was field-tested by EPA's Superfund Innovative Technology Evaluation (SITE) Program to evaluate the performance of the technology at the Bay City Middleground Landfill located in Bay City, Michigan (Ref. 14).

Table 3-5. Performance of GPCR™ Technology

Site	Location	Period	POP	Quantity of Soil Treated	Scale	Destruction Efficiency
Kwinana Commercial Operations	Australia	1995 to 2000	PCBs	2,000 tonnes (2,200 tons)	Full	> 99.9999%
			DDT			> 99.9999%
Kwinana Hex Waste Trials	Australia	April 1999	HCB	8 tonnes (9 tons)	Full	> 99.9999%
General Motors of Canada Limited	Canada	1996 to 1997	PCB	1,000 tonnes (1,100 tons)	Full	> 99.99999%
			Dioxins			> 99.9995%

Source: Ref. 12

The technology has been selected by the United Nations Industrial Development Organization for a pilot-scale project to treat approximately 1,000 tons of PCB-contaminated waste in Slovakia. The technology has also been licensed in Japan for treatment of PCB- and dioxin-contaminated wastes (Refs. 33 and 58).

The technology was developed by Eco Logic International Inc. in Ontario, Canada. Bennett Environmental Inc. in Oakville, Ontario, Canada, recently acquired exclusive patent rights to the technology. An update from previous reports is that this technology is not currently marketed, as it is considered to be cost prohibitive. However, Bennett Environmental Inc. is modifying the technology to improve its cost effectiveness (Ref. 40).

3.1.4 GeoMelt™

> **POPs TREATED: DIELDRIN, CHLORDANE, HEPTACHLOR, DDT, HCB, PCBS, DIOXINS, AND FURANS**
>
> **MEDIUM: SOIL AND SEDIMENTS**
>
> *FULL SCALE*
> *EX SITU AND IN SITU*

GeoMelt™ has been used to treat high-strength wastes containing POPs. The technology is available for both in situ and ex situ applications and in both fixed and transportable configurations. GeoMelt™ vitrification is a high-temperature technology that uses heat to destroy POPs and to permanently immobilize residual contaminants by incorporating them into the vitrified end product. GeoMelt's in situ process is available in two main configurations, In-Situ Vitrification (ISV) and Subsurface Planar Vitrification (SPV™). Both configurations use electrical current to heat, melt, and vitrify material in place. ISV is suitable for treatment to depths exceeding 10 feet. SPV is suitable for more shallow applications. GeoMelt™ also provides a variation of SPV called Deep-SPV which facilitates focused vitrification of limited-thickness treatment zones greater than 30 feet deep. Electric current is passed through soil using an array of electrodes

> **THE FACT SHEET PREPARED BY IHPA IS AVAILABLE AT HTTP://WWW.IHPA.INFO/LIBRARYNATO.HTM.**

inserted vertically into the surface of the contaminated zone. Because soil is not electrically conductive, a starter pattern of electrically conductive graphite and glass frit is placed in the soil between the electrodes. When power is fed to the electrodes, the graphite and glass frit conduct current through the soil, heating the surrounding area and melting directly adjacent soil. Once molten, the soil becomes conductive. The melting proceeds outward and downward. Typical operating temperatures range from 1,400 to 2,000 °C. As the temperatures increase, contaminants may begin to volatilize. When sufficiently high temperatures are attained, most organic contaminants are destroyed in situ through thermally mediated chemical reactions, yielding carbon dioxide, water vapor, and sometimes hydrogen chloride gas (if chlorinated contaminants are present). Gaseous reaction products (such as hydrogen chloride) and volatilized contaminants that escape in situ destruction are collected by an off-gas hood and are processed through an aboveground off-gas treatment system before their discharge to the atmosphere. When the heating stops, the medium cools to form a crystalline monolith vitrified end product, encapsulating contaminants that were not destroyed or volatilized (Ref. 26).

Geomelt ™ ICV process. Source: Ref. 24

GeoMelt's ex situ process, which is called In Container Vitrification (ICV™), involves heating contaminated material in a refractory-lined container. A hood placed over the container collects off-gases. The heat is generated by either two or four 12-inch-diameter, graphite electrodes positioned vertically in the container. Typical operating temperatures range from 1,400 to 2,000 °C. At these temperatures, the waste matrix melts, and organic contaminants are destroyed or volatilized. The off-gas from the process enters an off-gas treatment system, which includes a baghouse particulate filter, high-efficiency particulate air (HEPA) prefiltration, a NO_x (oxides of nitrogen) scrubber, a hydrosonic scrubber, a mist eliminator, a heater, and one or two HEPA filters. After treatment, the hood is removed and a lid is installed on the refractory-lined container. When the melt has solidified, the vitrified waste-filled container is disposed in a landfill based on the results of EPA Toxicity Characteristic Leaching Procedure (TCLP) analysis.

GeoMelt™ is a full-scale treatment technology and has been used to treat such POPs as dieldrin, chlordane, heptachlor, DDT, HCB, PCBs, dioxins, and furans (Ref. 26). GeoMelt™ has also been used to treat radioactive waste. Table 3-6 provides performance information for the technology.

Table 3-6. Performance of GeoMelt™ Technology

Site	Location	Period	POP	Quantity of Soil Treated	Scale	Untreated Concentration (mg/kg)	Treated Concentration (mg/kg)
Parsons Chemical/ ETM Enterprises Superfund Site	Grand Ledge, Michigan	1993 to 1994	DDT	4,350 tons	Full	340	<4
			Chlordane			89	<1
			Dieldrin			4.6	<0.08
TSCA Spokane	Spokane, Washington	1994 to 1996	PCBs	5,375 tons	Full	17,860	ND
Wasatch Chemical	Salt Lake City, Utah	1995 to 1996	Dioxins	5,440 tons	Full	0.011	ND
			DDT			1.091	ND
			Chlordane			535	ND
			HCB			17	<0.08
WCS-Commercial TSCA cleanup	Andrews, Texas	2005	PCBs	5 tons	Full	496	ND
WCS-Rocky Flats	Andrews, Texas	2005	PCBs	11 tons	Pilot	130	ND

Source: Ref. 24 and 33

Notes:
mg/kg = Milligram per kilogram
ND = Below detection limit
TSCA = Toxic Substance Control Act
WCS = Wasatch Chemical Superfund

GeoMelt™ is commercially available from AMEC Earth and Environmental, the sole licensee of this technology in the United States. AMEC Earth and Environmental owns several GeoMelt™ systems of varying sizes that are currently available for use.

3.1.5 In Situ Thermal Desorption

THE FACT SHEET PREPARED BY EPA IS INCLUDED IN APPENDIX C.

ISTD has been used to treat both high- and low-strength wastes containing POPs. ISTD is primarily an in situ technology but has also been used ex situ on constructed soil piles. ISTD is a thermally enhanced, in situ treatment technology that uses conductive heating to directly transfer heat to environmental media. "ISTD" has been a nonspecific term used to refer to in situ technologies that use heat to enhance the removal of volatile subsurface contaminants. The most common ISTD methods include steam and resistive heating. Enhanced soil vapor extraction (ESVE) is a commonly used synonym for ISTD. However, this section uses "ISTD" to refer to a specific proprietary technology that is distinct in its methods. ISTD, sometimes also known as "In Situ Thermal Destruction" is a patented technology developed by Shell Oil Co. and TerraTherm, which holds the exclusive license to the technology and is currently the only vendor.

There are three basic elements in the ISTD process (Ref. 51):

1. Application of heat to contaminated media by thermal conduction
2. Collection of desorbed contaminants through vapor extraction
3. Treatment of collected vapors

ISTD process at the Alhambra site. Source: Ref. 51

ISTD uses surface heating blankets or buried, electrically powered heaters to heat contaminated media. In the most common setup, a vertical array of heaters is placed in wells drilled into the remediation zone. Surface heating blankets are less commonly used. As the matrix is heated, adsorbed and liquid-phase contaminants begin to vaporize. Once high soil temperatures are achieved, a significant portion of the organic contaminants either oxidizes (if sufficient air is present) or pyrolizes. Desorbed contaminants are recovered through a network of vapor extraction wells. Contaminant vapors captured by the extraction wells are conveyed to an off-gas treatment system for treatment prior to their discharge to the atmosphere. TerraTherm offers two different methods of vapor treatment. One method treats extracted vapor without phase separation and uses a thermal oxidizer to break down organic vapors to primarily carbon dioxide and water. Thermal oxidation may be followed by vapor phase activated carbon absorption. The second method uses a heat exchanger to cool extracted vapors. The resulting liquid phase is then separated into aqueous and nonaqueous phases. The nonaqueous-phase liquid (NAPL) is usually disposed of at a licensed treatment, storage, or disposal facility. The aqueous phase is passed through liquid-phase activated carbon adsorption units and is then discharged. Cooled, uncondensed vapor is passed through vapor-phase activated carbon adsorption units and is then vented to the atmosphere (Refs. 6, 21 and 51).

POPs TREATED: PCBs, DIOXINS, AND FURANS

MEDIUM: SOIL AND SEDIMENT

COSTS: $200 TO $600 PER CUBIC YARD (COST IN USD, DATA FROM 1996 TO 2005)

FULL SCALE
IN SITU

Pilot- and full-scale applications of ISTD have been used to remove PCBs, dioxins, and furans. According to TerraTherm, laboratory-scale work indicates that this technology can also effectively treat other POPs, including aldrin, dieldrin, endrin, chlordane, heptachlor, DDT, mirex, HCB, and toxaphene. However, these contaminants have not yet been treated using ISTD at full or pilot scale. ISTD was field-tested by EPA's SITE Program to evaluate the performance of the technology at the Rocky Mountain Arsenal (RMA) site near Denver, Colorado. The site was contaminated with hexachlorocyclopentadiene, aldrin, chlordane, dieldrin, endrin, and isodrin (Ref. 20).

Four full-scale and two pilot-scale ISTD projects at POP-contaminated sites were identified. In general, treatment costs in USD at these sites ranged from $200 to $600 per cy. Projects involving ISTD treatment of larger volumes of waste will have lower unit costs. Available performance information for the technology is presented in Table 3-7.

Table 3-7. Performance of ISTD Technology

Site	Location	Period	POP	Quantity of Soil Treated	Scale	Untreated Concentration	Treated Concentration
Former South Glens Falls Dragstrip	Moreau, New York	1996	PCBs	NA	Full	5,000 mg/kg	0.8 mg/kg
Tanapag Village	Saipan, Northern Mariana Islands	July 1997 to August 1998	PCBs	1,000 cy	Full	10,000 mg/kg	< 1 mg/kg
Centerville Beach	Ferndale, California	September to December 1998	PCBs	667 cy	Full	860 mg/kg	< 0.17 mg/kg
			Dioxins and Furans			3.2 μg/kg	0.006 μg/kg
Missouri Electric Works	Cape Girardeau, Missouri	March to June 1997	PCBs	NA	Pilot	20,000 mg/kg	<0.033 mg/kg
Former Mare Island Naval Shipyard	Vallejo, California	September to December 1997	PCBs	222 cy	Pilot	2,200 mg/kg	<0.033 mg/kg
Alhambra "Wood Treater"	Alhambra, California	May 2002 to January 2005	Dioxins	16,200 cy	Full	194 μg/kg	<1 μg/kg

Source: Refs. 7, 48, 50 and 51

Notes:
cy = Cubic yard
μg/kg = Microgram per kilogram
mg/kg = Milligram per kilogram
NA = Not available

3.1.6   Mechanochemical Dehalogenation

MCD™ has been used to treat high-strength wastes containing POPs. The MCD™ technology uses mechanical energy to promote reductive dehalogenation of contaminants. In this process, contaminants react with a base metal and a hydrogen donor to generate reduced organics and metal salts. The base metal is typically an alkali-earth metal, an alkaline-earth metal, aluminum, zinc, or iron. The hydrogen donors used include alcohols, ethers, hydroxides, and hydrides. The process occurs ex situ in an enclosed ball mill, and the grinding medium provides the mechanical energy and mixing. The technology is applicable to soil, sediments, and mixed solid-liquid phases. The by-products generated at the end of the process are nonhazardous organics and metal salts (Ref. 54). Additional information about the technology is available at http://www.ihpa.info/libraryNATO.htm.

> POPs TREATED: DDT, ALDRIN, AND DIELDRIN
>
> MEDIUM: SOIL, SEDIMENT AND LIQUID WASTES
>
> *FULL SCALE*
> *EX SITU*

MCD process at the Mapua Site. Source: Ref. 54

One MCD™ process developed by Environmental Decontamination Ltd. (EDL) is being used at full scale to treat soil at the Fruitgrowers Chemical Company site in Mapua, New Zealand. The site is the location of a former pesticide and herbicide manufacturing plant that operated from 1950 to 1980. The site is approximately 8.4 acres in area and contains about 706,280 cubic feet of soil contaminated with DDT, DDD, DDE, aldrin, dieldrin, and lindane. Proof of performance testing of the MCD™ process was conducted at the site between February 16 and April 23, 2004. The objective of the testing was to demonstrate the technology's ability to treat the contaminated soil to meet the cleanup standards for commercial land use. The cleanup criteria are listed in Table 3-8 and the proof of performance testing results are listed in

Table 3-9. The criteria are based on the concentration of DDX (the sum of the concentrations of DDT, dichlorodiphenyldichloroethane [DDD], and dichlorodiphenyldichloroethylene [DDE]) and the sum of the concentrations of aldrin, dieldrin, and lindane.

THE FACT SHEET PREPARED BY IHPA IS AVAILABLE AT HTTP://WWW.IHPA.INFO/LIBRARYNATO.HTM.

Table 3-8. Soil Acceptance Criteria for the Mapua Site

Land Use	Depth (meters)	DDX (Total DDT, DDD, and DDE) (mg/kg)	Aldrin + Dieldrin + Lindane (mg/kg)
Commercial	0 to 0.5	5	3
	' Below 0.5	200	60

Source: Ref. 54

Notes:
mg/kg = Milligram per kilogram

At the Mapua site, soil greater than 10 millimeters (mm) size fraction has contaminant concentrations below the soil acceptance criteria for the site and requires no treatment. EDL receives contaminated soil which is less than 10 mm in size. The 10 mm size fraction soil is dried and then passed through a 2 mm screen to segregate soil particles less than and greater than 2 mm size. Contaminated soil less than 2-mm size is treated using the MCD™ process. Additional information on the soil drying and screening processes and the MCD™ process are described below.

Soil Drying
The contaminated soil enters a temperature controlled, diesel-fired rotary drum unit. As the soil passes through the drier, the soil particles undergo size reduction. The moisture content in soil exiting the drier is typically less than 2 percent. Gaseous emissions from the drier are treated by an air quality control system consisting of cyclones, a baghouse, a scrubber and an activated carbon filter.

Soil Screening

Soil exiting the drier is passed through a rotary screen to separate soil particles by size range. Soil particles less than 2 mm in size are separated from soil particles between 2 and 10 mm in size. Soil samples are collected from the 2- to 10-mm fraction stream and analyzed. Thus far, DDX concentrations in this size fraction have been at or below cleanup standards and have consequently not required treatment. The less than 2-mm fraction stream and the fines from the cyclones and baghouse are fed into the MCD™ reactor.

MCD™ Process

The dried contaminated soil (the less than 2-mm fraction) and the fines from the cyclones and baghouse fed into the MCD™ reactor are mixed with metered quantities of a combination of metal salts and a hydrogen donor at a rate of around 3 percent by mass. The reactor is a vibratory mill that has two horizontally mounted cylinders containing a grinding medium. The grinding medium provide the mechanical impact energy required to drive the chemical reaction. Treated soil exits the base of the MCD™ reactor through enclosed screw conveyors and enters a paddle mixer, where the treated material is wetted to minimize dust generation. The required residence time within the reactor is about 15 minutes. The treated soil is then analyzed. Once treatment of soil to cleanup standards has been completed, treated soil is placed in a clean backfill area.

During the proof of performance testing at the Fruitgrowers Chemical Company site in Mapua, New Zealand, the MCD™ system exhibited a maximum treatment rate of 139 cubic meters per week. Table 3-9 lists the initial and final mean contaminant concentrations in the soil treated in the MCD™ reactor. The concentrations listed in Table 3-9 are mean concentrations in samples collected between February 16 and April 23, 2004. The treated soil met the cleanup criteria for soil more than 0.5 m below ground surface but did not meet the criteria for soil from 0 to 0.5 m below ground surface.

Table 3-9. Performance of MCD™ Technology at the Mapua Site

POP	Untreated Concentration (mg/kg)	Treated Concentration (mg/kg)	Percent Reduction	Soil Acceptance Criteria by Depth (meters)	
				0 to 0.5	> 0.5
DDX	717	64.8	91%	5	200
Aldrin	7.52	0.798	89%	NA	NA
Dieldrin	65.6	19.8	70%	NA	NA
Lindane	1.25	0.145	88%	NA	NA
Aldrin+Dieldrin +Lindane	73.245	20.612	72%	3	60

Source: Ref. 54

Notes:
mg/kg = Milligram per kilogram
NA = Not available

Subsequent to the Proof of Performance testing, EDL was commissioned to remediate the site with an expected completion date in 2006. EDL has developed a proprietary reactor that eliminates the need for a vibratory ball mill. Experience has indicated that soil composition affects the performance of the process. Clays in particular have been shown to have a negative performance impact. EDL plans to conduct pilot tests on its reactor in the United States during the later part of 2005. These trials will involve DDT, lindane, and PCBs.

The MCD™ technology is available from EDL in Auckland, New Zealand (http://edl.net.nz/about.php), and from Tribochem in Wunstrof, Germany (http://www.tribochem.com) (Ref. 8). Information was provided by EDL. Tribochem has not provided process details, performance data, or costs for its technology.

3.1.7 Xenorem™

Xenorem™ is an ex situ bioremediation technology that has been used to treat low-strength wastes containing chlordane, DDT, dieldrin, and toxaphene contamination. Xenorem™ uses an enhanced composting technology consisting of aerobic and anaerobic treatment cycles. Organic amendments such as manure and wood chips are added to contaminated soil, which can increase the final amended soil volume by as much as 40 percent (Ref. 30). A self-propelled SCAT windrow incorporates the amendments into the soil and provides aeration creating aerobic conditions. The presence of high levels of available nutrients from the amendment increases the metabolic activity in the amended soil and depletes the oxygen content, creating anaerobic conditions. The anaerobic conditions promote dechlorination of organochlorine compounds. The length of the anaerobic phase is determined

> **POPS TREATED: CHLORDANE, DDT, DIELDRIN, AND TOXAPHENE**
>
> **MEDIUM: SOIL**
>
> **COSTS: $132 PER CUBIC YARD (COST IN 2000 USD)**
>
> *FULL SCALE*
> *EX SITU*

by bench-scale studies. At the end of the anaerobic phase, the SCAT unit is used to mix the amended soil, creating aerobic conditions again. The anaerobic and aerobic cycles are repeated until the desired contaminant reductions are achieved. Typically, by the end of 14 weeks of treatment the organic amendments are spent. Soil samples are collected from the treated soil, and if the contaminant concentrations do not meet the cleanup goals, more organic amendments are added; the treatment is continued as long as necessary.

This technology was applied in a full-scale cleanup at the Stauffer Management Company Superfund site in Tampa, Florida. The site is the location of a pesticide manufacturing and distribution facility that operated from 1951 to 1986 (Ref. 13). Soil on the 40-acre site was contaminated with chlordane, DDD, DDE, DDT, dieldrin, molinate, and toxaphene. The Xenorem™ technology was applied to two 4,000-cy batches of soil. The first batch was completed in 2001 and the second batch was completed in 2002. The contaminated soil was excavated; screened; mixed; and amended with dairy cow manure, chicken litter, and wood chips. The amended soil matrix was then placed in a compost windrow. The temperature, oxidation-reduction potential, and moisture level of the amended soil matrix were continuously monitored (Ref. 13). Table 3-10 presents the performance data for Batch 1 and Batch 2. Batch 1 was treated for a total of 24 weeks and achieved the site cleanup goals for chlordane, DDD, DDE, dieldrin, and molinate. After 12 weeks of treatment, Batch 2 achieved the site cleanup goals for chlordane, DDE, dieldrin, and molinate. The treatment of Batch 2 extended beyond 12 weeks; the final performance data for Batch 2 are not yet available from the vendor. Neither batch achieved the site cleanup goals for DDT and toxaphene. Typical treatment costs in USD using Xenorem™ were provided by the vendor and are approximately $132 per cy of contaminated soil (Ref. 16)

The Xenorem™ technology was applied to a third batch of contaminated site soil. Batch 3 was treated for one year but did not achieve the cleanup goals for chlordane, DDT, dieldrin, and toxaphene. Because the selected remedy did not fully meet the cleanup goals, the remedial design for the site is being modified. EPA is awaiting details of the modification proposal. Eventually EPA will prepare an Explanation of Significant Difference (ESD) fact sheet explaining the selection of a new remedy (Ref. 29).

Table 3-10. Performance of Xenorem™ Technology at the Tampa Site

Pesticide	Site Cleanup Goal (mg/kg)	Batch 1 [a]			Batch 2 [b]		
		Untreated Concentration (mg/kg)	Treated Concentration (mg/kg)	Percent Reduction	Untreated Concentration (mg/kg)	Treated Concentration (mg/kg)	Percent Reduction
Chlordane	2.3	3.8	< MDL	NA	4.5	1.2	75%
DDD	12.6	26	9.3	65%	24	14	42%
DDE	8.91	6.6	2.1	68%	6.1	2.6	57%
DDT	8.91	82	9.8	88%	196	14	93%
Dieldrin	0.19	2.4	<MDL	NA	2.7	0.7	74%
Molinate	0.74	0.2	<MDL	NA	0.4	<MDL	NA
Toxaphene	2.75	129	7.8	94%	139	23	83%

Source: Ref. 30

Notes:

MDL = Method detection limit (the MDL was not provided in the source document)

mg/kg = Milligram per kilogram

NA = Not available

[a] For Batch 1, treated concentrations are at the end of a 24-week period.

[b] For Batch 2, treated concentrations are at the end of a 12-week period.

Quantity treated: 4,000-cy of soil (Batch 1 and Batch 2).

Xenorem™ is a patented technology developed by Stauffer Management Company, a subsidiary of AstraZeneca Group PLC in Mississauga, Ontario, Canada. Recently, this technology was sold to the University of Delaware (Ref. 29). Additional information on the technology can be obtained from the Technology Transfer Corporation at the University of Delaware in Newark, Delaware.

3.2 Pilot-Scale Technologies for Treatment of POPs

This section describes technologies that have been implemented to treat POPs at the pilot scale. Each subsection focuses on a single technology and includes a description of the technology and information about its application at specific sites. Fact sheets developed by IHPA contain additional details on some of these technologies and their applications. Links to the IHPA fact sheets are included in the appropriate subsections of this report.

3.2.1 Base-Catalyzed Decomposition

BCD is an ex situ technology that has been used in pilot tests to treat high-strength soil containing POP contamination. The technology is available in both transportable and fixed configurations.

The BCD technology uses a two-stage process. In the first stage of the treatment process, contaminated soil is mixed with an alkali such as sodium bicarbonate, and the mixture is heated in a thermal desorption reactor to temperatures ranging from 315 to 500 °C. The heat separates the halogenated compounds from the soil by evaporation. In the second stage of the process, the volatilized contaminants pass through a condenser. The condensate is then sent to a BCD liquid tank reactor (LTR). Sodium hydroxide, a proprietary catalyst, and carrier oil are added to the LTR, which is then heated to above 326 °C for 3 to 6 hours. The carrier oil serves both as a suspension medium and a hydrogen donor. The heated oil is then cooled and sampled to determine whether it meets disposal criteria. If the oil does not meet the disposal criteria, it is returned to the LTR, reagents are added, and the reactor is reheated (Ref. 37). The treated soil can be used as backfill on site.

At the Warren County PCB Landfill site in Warren County, North Carolina, a full-scale demonstration of BCD was proposed by the North Carolina Department of Environment and Natural Resources (NCDENR). After completing the first stage,

<table>
<tr><td>THE FACT SHEET PREPARED BY IHPA IS AVAILABLE AT HTTP://WWW.IHPA.INFO/LIBRARYNATO.HTM.</td></tr>
</table>

thermal desorption of contaminated site soil, NCDENR decided to incinerate the residual, highly contaminated waste oil condensate instead of performing the second-stage BCD reaction (Ref. 34). The off-site incineration activities were completed in June 1996.

<table>
<tr><td>
POPs TREATED: PCBs, DIOXINS, AND FURANS

PRETREATMENT: THERMAL DESORPTION

MEDIUM: SOIL AND LIQUIDS

PILOT SCALE
EX SITU
</td></tr>
</table>

BCD has been implemented at a bench scale to treat soil contaminated with POPs. In 2004, EPA sent a sample of contaminated waste oil from the Warren County PCB Landfill site to an analytical laboratory in order to perform a bench-scale demonstration. Results from the bench scale study indicate that total PCBs were reduced from 81,100 mg/kg to below the detection limit of 5 mg/kg. Total tetrachlorodibenzodioxin (TCDD) was reduced from 5,800 nanogram per kilogram (ng/kg) to 9.1 ng/kg. A second bench scale study was conducted by EPA in 2005. The second bench scale study indicated that PCBs were reduced from 5,280 mg/kg to below the detection limit of 5 mg/kg. Total TCDD was reduced from 5,800 ng/kg to 15.0 ng/kg (Ref. 37).

BCD was developed by EPA's National Risk Management Research Laboratory in Cincinnati, Ohio. EPA holds the patent rights to this technology in the United States. The foreign rights for this technology are held by BCD Group Inc., Cincinnati, Ohio. The technology has been licensed by BCD Group Inc., to environmental firms in Spain, Australia, Japan and Mexico. Since the invention of the BCD technology in 1990, considerable technology advancements have been made with the discovery of a new catalyst. The catalyst used in the second generation BCD technology reduces the reaction time in the BCD reactor (Ref. 44). This second generation technology has been in Australia, Mexico and Spain to treat PCB contaminated oil. Two commercial BCD plants are being constructed in Czech Republic and will begin operation in 2006. At this time, performance data for the BCD operations in Australia, Mexico, and Spain are not available from the vendors.

3.2.2 CerOx™

CerOx™ is an ex situ electrochemical reaction technology that has been used in pilot tests to treat low-strength liquids

<table>
<tr><td>
POPs TREATED: CHLORDANE, DIOXINS, AND PCBs

PRETREATMENT: SOIL AND SEDIMENT ARE MIXED WITH WATER TO PRODUCE A FLUID INFLUENT

MEDIUM: LIQUIDS

PILOT SCALE
EX SITU
</td></tr>
</table>

containing POP contamination. CerOx™ uses cerium in its highest valence state (IV) to oxidize organic compounds, including POPs, to form carbon dioxide,

water, and inorganic acid gases. The technology uses an electrochemical cell to produce cerium (IV) from cerium

CerOx™ treatment system,
Source: Ref. 9

(III). Prior to treatment, solid waste such as soil or sediment is mixed with water to produce a fluid waste stream. This waste stream is injected with cerium (IV) from the electrochemical cell, agitated through sonication, and transferred to a liquid-phase reactor. The liquid-phase reaction takes place at a temperature between 90 and 95 $^{\circ}$C and results in the destruction of organic compounds in the waste stream. During this process, cerium (IV) is reduced to cerium (III). Cerium (III) and unreacted cerium (IV) are returned to the electrochemical cell for recycling, and the treated medium is removed from the system. Gases produced during the liquid-phase reaction usually include carbon dioxide, chlorine gas, and unreacted volatile organic compounds Volatile Organic Compounds (VOC). These gases are processed through a gaseous-phase reactor that uses cerium (IV) to destroy VOCs. The remaining gases are passed through a scrubber to remove acid gases and are then vented to the atmosphere. Liquid from the scrubber is discharged (Ref. 9).

The information sources used to prepare this report did not describe any applications of CerOx^{TM} systems at a pilot or full scale for treatment of POP-contaminated soil or sediment. CerOx™ systems have been used to treat POP-contaminated liquids. The first CerOx™ system was installed at the University of Nevada at Reno (UNR) to destroy surplus chlorinated pesticides and herbicides from the university's agricultural departments. Prior to use of this system by UNR, CerOx Corporation conducted proof of performance tests in May 2000. The medium treated was a pesticide-water emulsion. In one test, 71 percent by mass chlordane was mixed with water and fed to the system. The system is reported to have achieved a chlordane destruction efficiency of 99.995 percent in the gaseous-phase reactor (Ref. 4). Chlordane concentrations in the liquid effluent were not reported.

> **THE FACT SHEET PREPARED BY IHPA IS AVAILABLE AT** HTTP://WWW.IHPA.INFO/LIBRARYNATO.HTM.

The vendor later performed additional tests of the UNR system to determine the ability of CerOx™ to treat PCBs and dioxins (Ref. 58). A treatment test was performed on August 29, 2000, using a feed stream consisting of three commercially available dioxins dissolved in isopropyl alcohol. The dioxins in the feed stream were present at a concentration of 5 parts per billion (ppb). Two of three samples collected from the system's effluent contained dioxins at concentrations lower than their detection limit of 0.397 part per trillion (ppt). One sample had a dioxin concentration of 0.432 ppt. The UNR system was tested again on August 30, 2000, using a liquid sample from a remedial operation being performed in Fayetteville, North Carolina. The sample consisted of an isopropyl alcohol solution containing about 2 parts per million (ppm) PCBs. The system effluent contained PCB concentrations less than the minimum detection limit of 0.4 ppb PCBs (Ref. 9).

The technology was developed by CerOx^{TM} Corporation in Santa Maria, California. CerOx^{TM} Corporation offers a variety of CerOx^{TM} treatment systems for commercial use. The systems range in size from modules with 25-gallon per day (gpd) treatment capacities to multimodular plants with 100,000-gpd treatment capacities (Ref. 9).

3.2.3 Phytotechnology

Phytotechnology is a process that uses plants to remove, transfer, stabilize, or destroy contaminants in soil, sediment, and groundwater. It may be applied in situ or ex situ to treat low-strength soils, sludges, and sediments contaminated with POPs. The mechanisms include:

> **MEDIUM: SOIL AND SEDIMENT**
>
> *PILOT SCALE*
> *EX SITU AND IN SITU*

- Enhanced rhizosphere biodegradation (degradation in the soil immediately surrounding plant roots),

- Phytovolatilization (the transfer of the pollutants to air via the plant transpiration stream),
- Phytoextraction (also known as phytoaccumulation, the uptake of contaminants by plant roots and the translocation/accumulation of contaminants into plant shoots and leaves),
- Phytodegradation (metabolism of contaminants within plant tissues),
- Phytostabilization (production of chemical compounds by plants to immobilize contaminants at the interface of roots and soil), and
- Hydraulic control (the use of trees to intercept and transpire large quantities of groundwater or surface water for plume control).

In general more proven field studies have been conducted using phytostabilization and hydraulic control mechanisms. Other proven uses of phytotechnologies include alternative landfill caps, the use of wetlands to improve water quality, and treatment of certain contaminants (such as petroleum products and chlorinated solvents).

Phytoremediation of POPs is not feasible for stockpiles of contamination but provides an appropriate polishing technology for residual contamination in soils. Initial laboratory research identified enhanced degradation of PCBs in the rhizosphere (Refs. 11, 27, and 36). Other researchers are finding promising results for phytoextraction in the laboratory and at the pilot scale phase. The Connecticut Agricultural Experimental Station's preliminary data has shown that a narrow range of plant species (certain cucurbitas) can effectively accumulate significant amounts of highly weathered pesticide residues such as DDE and chlordane from soil (Ref. 61). The Royal Military College of Canada has also demonstrated that certain plants species can extract and store significant levels of PCBs and DDT (Ref. 62). Both the Ukraine and Kazakhstan have been conducting research on the use of plants to remediate soils laced with pesticides. In the Ukraine, laboratory experiments have shown that bean plants can accumulate and decompose DDT (Ref. 38). In Kazakhstan, native vegetation that can tolerate and accumulate pesticides has been identified (Ref. 39).

While research is still active and needed, field scale projects are also occurring. A clean-up project was conducted at a 40 year old scrap yard site with PCB contaminated soils at the 225 ppm level. The site contamination was approximately 2 acres in area and three feet deep. The clean-up project demonstrated that PCB concentrations decreased (over 90%) in the presence of red mulberry trees and bermuda grasses within 2 years (Ref. 31). Another example is an Evapotranspiration Cover that will be constructed at the Rocky Mountain Arsenal (RMA) National Wildlife Refuge near Denver, Colorado; the cover will address contaminants including aldrin, chlordane, DDT, dieldrin, and endrin. A field demonstration project was used as the basis of the final design, which will include five projects that encompass 400 acres. The selected seed mix for the site includes ten grass species and ten wildflower species native to the site (Ref. 34).

Furthermore, two EPA Superfund sites have utilized phytotechnology as a treatment for POPs:

- Aberdeen Pesticides Dumps in North Carolina is utilizing phytotechnology for residual contaminants (dieldrin and HCB) using poplar trees and grasses. This is an on-going project.
- Fort Wainwright in Alaska utilized ex-situ phytotechnology for aldrin and dieldrin with willow trees. After treatment, the soil was deposited in the site landfill rather than a hazardous waste landfill.

3.2.4 Solvated Electron Technology

Solvated electron technology uses solvated electron solutions to reduce organic compounds to metal salts and the parent dehalogenated molecules. Solvated electron solutions are formed by dissolving alkali or alkaline earth metals such as sodium, calcium, and lithium in solvents such as anhydrous liquid ammonia.

> **THE FACT SHEET PREPARED BY IHPA IS AVAILABLE AT** HTTP://WWW.IHPA.INFO/LIBRARYNATO.HTM.

Commodore Solution Technologies, Inc., the vendor of the technology, is not currently marketing the technology because of its high treatment costs.

3.2.5 Sonic Technology

Sonic technology is an ex situ technology that is used to treat low- and high-strength soils containing PCB contamination.

> **POPs TREATED: PCBs**
>
> **PRETREATMENT: SOIL IS MIXED WITH SOLVENT TO PRODUCE A SLURRY**
>
> **MEDIUM: SOIL**
>
> *PILOT SCALE*
> *EX SITU*

In this process, contaminated soil is first mixed with a solvent. The mixture is then subjected to sonic energy generated by a proprietary low-frequency generator. Using sonic energy, the mixture is agitated and the PCBs from the soil are extracted and suspended in the solvent. The solvent is then separated from the mixture using multistage liquid separators. The solvent is then mixed with elemental sodium, and subjected to sonic energy again. The sonic energy activates dechlorination of the PCBs in the solvent. The spent solvent can then be recycled through the system. Any off-gas from the process is treated using condensation, demisting, and multistage carbon filtration (Ref. 45).

In a pilot scale application of the technology to treat PCB-contaminated soil, the concentrations of PCBs before treatment were 388 to 436 mg/kg, and the concentrations after treatment were 0.35 to 0.81 mg/kg. The technology is being implemented at full scale to treat approximately 3,000 tons of PCB-contaminated soil at the Juker Holdings site in Vancouver, British Columbia, Canada (Ref. 45). Additional performance data of the full scale application of this technology are not currently available.

The technology was developed by Sonic Environmental Solutions Inc. in Vancouver, Canada.

Sonic technology process. Source: Ref. 45

3.3   Bench-Scale Technologies for Treatment of POPs

This section describes the technologies that have been implemented to treat POPs at the bench scale. Each subsection focuses on a single technology and includes a description of the technology and information about its application at specific sites. Fact sheets developed by IHPA contain additional details on some of these technologies and their applications. Links to the IHPA fact sheets are included in the appropriate subsections of this report.

3.3.1   Self-Propagating High-Temperature Dehalogenation

Self-Propagating High-Temperature Dehalogenation (SPHTD) is an ex situ technology used to treat stockpiles containing HCB contamination.

HCB containing stockpiles are mixed with calcium hydride or calcium metal, and the mixture is placed in a reaction chamber containing a tungsten coil. Addition of purified argon gas causes the reaction chamber to become pressurized, and an electrical

> **THE FACT SHEET PREPARED BY IHPA IS AVAILABLE AT HTTP://WWW.IHPA.INFO/LIBRARYNATO.HTM.**

pulse to the tungsten coil initiates the reaction. Once initiated, the reductive reactions that occur in the reaction chamber are exothermic and self-propagating. The reaction chamber can reach a temperature of 3,727 °C, which creates thermochemical conditions that convert HCB to calcium chloride, carbon, and hydrogen (Ref. 58).

> **POPs TREATED: HCB**
>
> **MEDIUM: POP STOCKPILES**
>
> *BENCH SCALE*
> *EX SITU*

SPHTD has been tested at bench scale using materials contaminated with HCB, but no bench-scale test results are available (Ref. 32). The information sources used to prepare this report did not provide information about application of the technology at the pilot or full scale. The SPHTD technology is being developed by Centro Studi Sulle Reazioni Autopropaganti in Italy.

3.3.2   TDR-3R™

TDR-3R™ is an ex situ technology used to treat high- and low-strength soils containing HCB contamination.

> **THE FACT SHEET PREPARED BY IHPA IS AVAILABLE AT HTTP://WWW.IHPA.INFO/LIBRARYNATO.HTM.**

The TDR-3R™ technology uses a continuous low-temperature thermal desorption process conducted in the absence of air. The main component of this process is a specially designed, indirectly fired, horizontally arranged rotary kiln. Contaminated soil is heated in the kiln to a temperature typically between 300 and 350 °C under an applied vacuum of 0 to 50 Pascal. In some instances, the kiln is heated to higher temperatures when POPs are being treated. The contaminants in the soil desorb and vaporize in the kiln. The vaporized contaminants are recovered from the kiln and combusted in a thermal oxidizer for at least 2 seconds at a temperature exceeding 1,250 °C. Off-gas from the thermal oxidizer is rapidly cooled, passed through a wet gas multi-venturi scrubber, and

> **POPs TREATED: HCB**
>
> **PRETREATMENT: THERMAL DESORPTION**
>
> **MEDIUM: SOIL**
>
> *BENCH SCALE*
> *EX SITU*

discharged. Process water from the scrubber is treated and discharged. Treated soil exiting the kiln is cooled indirectly and removed (Refs. 33 and 52).

TDR-3R™ has been implemented at a bench scale in Gare, Hungary, to treat 100 kilograms (kg) of soil contaminated with HCB. Treatment occurred at a temperature of 450 °C under a vacuum of 30 Pascal.

The technology reduced the soil's HCB concentration from 1,215 to 0.1 mg/kg (Ref. 53).

TDR-3R™ is marketed by Terra Humana Clean Technology Engineering Ltd. in Hungary. This firm is a subsidiary of Thermal Desorption Technology Group LLC in the United States. The firm has developed pilot scale kilns that operate at a throughput of 0.1 tons per hour. Larger kilns that operate at throughputs from 5 to 70 tons per hour are still conceptual (Ref. 52).

3.3.3 Mediated Electrochemical Oxidation (AEA Silver II™)

AEA Silver II™ is an ex situ technology used to treat low-strength wastes containing POPs.

The AEA Silver II™ process uses Ag^{2+} ions to oxidize organic compounds, including POPs, to form carbon dioxide, neutral salts, and dilute acid solutions. The process operates at low temperatures (60 to 90 °C) and at atmospheric pressure.

> **THE FACT SHEET PREPARED BY IHPA IS AVAILABLE AT** HTTP://WWW.IHPA.INFO/LIBRARYNATO.HTM.

According to AEA Technologies Inc., the vendor of the technology, AEA Silver II™ is not applicable for soil or sediment. The contaminant has to be in an aqueous phase for the technology to be applied. Therefore, pretreatment is needed to extract the contaminant from the solid phase to an aqueous phase.

AEA Technologies Inc. is not currently marketing this technology.

3.4 Full-Scale Technologies with Potential to Treat POPs

This section describes technologies that have been implemented to treat non-POPs at full scale and that are potentially applicable for treatment of POPs. Each subsection focuses on a single technology and includes a description of the technology and information about its application at specific sites. Fact sheets developed by IHPA contain additional details on some of these technologies and their applications. Links to the specific IHPA fact sheets are included in the appropriate subsections of this report.

3.4.1 Plasma Arc

Plasma arc technologies use a thermal plasma field to treat contaminated wastes. The plasma field is created by directing electric current through a gas stream under low pressure to form a plasma with a temperature ranging from 1,600 to 20,000 °C. Bringing the plasma into contact with the waste causes contaminants to dissociate into their atomic elements. The separated elements are subsequently cooled, which causes them to recombine to form inert compounds. The process may also destroy organic compounds through pyrolysis. The end products are typically gases, such as carbon monoxide, carbon dioxide, hydrogen and inert solids. If chlorinated compounds are present in the waste, acid gas is also generated as an end product. The off-gas from the plasma arc system passes through an off-gas treatment system and is then discharged. The plasma arc technologies that are used to treat organic wastes include PLASCON™, Plasma Arc Centrifugal Treatment (PACT), and the Plasma Converter System (PCS). The PLASCON and PCS may potentially remediate POPs; however, the PACT technology has treated POP contamination at pilot scale. These technologies are described below (Ref. 58).

3.4.1.1 PLASCON™

PLASCON™ is an ex situ technology that can be used to treat both solid and liquid waste streams. It is potentially applicable to both low- and high-strength wastes containing POP contamination.

The PLASCON™ technology passes a direct current discharge through argon gas to create plasma with a temperature greater than 10,000 °C. Liquid or gaseous waste is injected directly into the plasma. Solid waste is pretreated using thermal desorption to extract volatile

THE FACT SHEET PREPARED BY IHPA IS AVAILABLE AT HTTP://WWW.IHPA.INFO/LIBRARYNATO.HTM.

contaminants. The extracted vapors are then condensed and injected into the plasma as liquid waste. Liquid waste is vaporized by heat transfer from the plasma. Organic compounds present in the waste pyrolize. The products formed during pyrolysis pass through a reaction tube providing sufficient residence time to ensure complete decomposition of the feed material. Gases exit the tube at a temperature of about 1,500 °C and are rapidly cooled to less than 100 °C in a spray condenser using an alkaline spray solution. The gases are further cooled and scrubbed of any remaining acid gases in a packed tower. Off-gases, which contain mainly carbon monoxide and argon, are then thermally oxidized to convert carbon monoxide to carbon dioxide (Ref. 33).

BCD Technologies Private Limited in Brisbane, Australia, purchased a PLASCON™ plant to treat PCB-contaminated wastes. This firm used a thermal desorption system in conjunction with PLASCON™ to treat a range of solid and semi-solid waste streams (Ref. 46). Performance data for this system are not available in the information sources used to prepare this report.

PLASCON™ has been used at full scale to treat various organic contaminants. SRL Plasma Pty. Ltd., an Australian company, is the patent holder of this technology.

3.4.1.2 Plasma Arc Centrifugal Treatment

PACT is an ex situ technology that can be used to treat low- and high-strength wastes containing POP contamination.

PACT uses heat generated by a plasma torch to melt and vitrify contaminated feed material. Primary treatment occurs inside a centrifuge tank housing the plasma torch. Centrifugal force produced by the rotating tank pushes the waste material away from the center and into the plasma torch's field of influence. The plasma torch heats waste material within its field of influence to a temperature of about 1,650 °C (Ref. 10). At the end of primary treatment, the tank stops rotating and the molten waste exits the tank through a chute at the center of the tank. Molten waste is collected in molds and cooled to form vitreous solids. Volatilized contaminants pass from the centrifuge tank to a natural gas-fueled secondary treatment tank. Secondary treatment of gaseous contaminants occurs at a temperature of about 1,000 °C. This part of the process ensures destruction of products of incomplete combustion such as dioxins and furans. Exhaust gases are discharged to an off-gas treatment system that cools the exhaust and scrubs it to remove acid gases (Ref. 43).

PACT has been used at a pilot scale to treat waste contaminated with HCB and has been used at full scale to treat contaminants other than POPs. Waste containing HCB was treated in a PACT demonstration plant in 1991 (Ref. 43). The vendor, Retech Systems LLC, plans to ship a PACT system to Russia in 2005 for treatment of capacitors contaminated with PCBs (Ref. 43).

PACT is a full-scale treatment technology that is manufactured and marketed in the United States by Retech Systems LLC.

3.4.1.3 Plasma Converter System

PCS is an ex situ technology that can be used to treat soil, liquid, and gaseous waste streams. It is potentially applicable to both low- and high-strength wastes containing POP contamination.

PCS uses plasma generated by a torch inside a cylindrical reaction chamber. Mixed waste fed into the reaction chamber passes through the plasma as the waste moves from one end of the chamber to the other. The arc inside the plasma can reach a temperature as high as 16,000 °C. Contaminants dissociate into their constituent elements within the plasma. The elements recombine outside the plasma to form gaseous and molten products. Molten material formed in the reaction chamber is removed from the bottom of the chamber and cooled to form inert solids. These solids can include metals and inert silicate stones. Exhaust gases from the reaction chamber pass through an off-gas treatment system and are then discharged. The off-gas treatment system includes a cyclonic separator for removal of particulate matter, cartridge filters for dust removal, a catalytic converter for reduction of oxides of nitrogen, and a scrubber for removal of acid gases. The recovered gas may be used to produce polymers or fuel gas (Refs. 10 and 47).

The PCS is a full-scale technology that is manufactured and marketed in the United States by STARTECH Environmental Corp. The plasma converter vessel is available in several sizes with treatment capacities ranging from 5 to 100 tons per day (Ref. 47). The technology vendor, STARTECH Environmental Corp., has stated that the PCS could be tailored to treat PCBs and HCB.

3.4.2 Supercritical Water Oxidation

SCWO is an ex situ technology that has been used to treat solid and liquid wastes. It is potentially applicable to both low- and high-strength wastes containing POP contamination. Current SCWO systems are non transportable.

SCWO occurs in an enclosed system at a temperature and pressure above the critical point of water (374 °C and 22.1×10^6 Pascal). Under these conditions, the gas-liquid phase boundary ceases to exist and water exists in a fluid state that is neither liquid nor gas.

THE FACT SHEET PREPARED BY IHPA IS AVAILABLE AT HTTP://WWW.IHPA.INFO/LIBRARYNATO.HTM.

Organic compounds have a higher solubility in supercritical water. An added oxidant such as oxygen or hydrogen peroxide reacts with dissolved organic contaminants in the supercritical water to form carbon dioxide, water, inorganic acids, and salts (Refs. 17 and 35).

The specifics of SCWO system design and operation vary. In general, currently available SCWO systems operate continuously, use corrosion-resistant materials in their reactors and process only fluid influents. One such system marketed by Turbosystems Engineering Inc. blends a contaminated aqueous stream with an oxidant from a storage tank. The blended stream is pressurized, preheated, and passed into the SCWO reactor. Contaminants are destroyed inside the reactor, and the effluent is cooled, depressurized, separated into liquid and gas streams, and discharged. SCWO technology is also available from General Atomics' Advanced Process Systems division (Ref. 25).

The Assembled Chemical Weapons Assessment (ACWA) Program was established in 1997 to test and demonstrate at least two alternative technologies to the baseline incineration process for the demilitarization of assembled chemical weapons (Ref. 5). In 2003, the Bechtel Parsons Blue Grass Team was awarded a contract to design, construct, test, operate, and close the Blue Grass Army Depot Destruction Pilot Plant using SCWO. The SCWO system is currently in the design phase. SCWO was also selected for use at the Newport Army Depot to destroy 1,269 tons of liquid agent VX. Existing SCWO systems are limited to the treating of liquids and solids with a particle size of less than 200 microns suspended in a liquid. The process is best suited to wastes with less than 20 percent organic content (Ref. 33). SCWO treatment of solid wastes after they have been ground into a fine slurry has been demonstrated using feed materials containing up to 25 percent suspended solids (Refs. 5, 28, 33, 35 and 58).

In the United States, SCWO technology is available from General Atomics' Advanced Process Systems division (Ref. 25). Although General Atomics primarily markets this technology to government clients, the firm has also designed and fabricated SCWO systems for commercial entities (Refs. 25 and 35). The SCWO process developed by General Atomics was selected for use as an Assembled Chemical Weapons Assessment technology to treat non-POPs such as GB, VX, H, HD, and TNT. Turbosystems Engineering Inc. also designs and markets SCWO systems in the United States (Ref. 55). Turbosystems Engineering Inc. claims that its system can treat DDT and HCB. No performance data substantiating this claim are available in the information sources used to prepare this report.

4.0   INFORMATION SOURCES

The following web based information sources were used during the preparation of this report. Additional information on POPs can be obtained from the web sites identified below as well as from the references listed in Section 6.0.

Stockholm Convention
http://www.pops.int/

International HCH and Pesticides Association
http://www.ihpa.info/libraryNATO.htm

Science and Technology Advisory Panel of the Global Environmental Facility
http://www.unep.org/stapgef/home/index.htm

U.S. Environmental Protection Agency
http://www.clu-in.org/POPs
http://www.clu-in.org/acwaatap
http://www.epa.gov/oppfod01/international/pops.htm

United Nations
http://www.basel.int
http://www.chem.unep.ch/pops/
http://www.unep.org/stapgef/documents/popsJapan2003.htm
http://www.gpa.unep.org/pollute/organic.htm
http://www.who.int/iomc/groups/pop/en/
http://www.unido.org/doc/29487

Other Sources
http://www.africastockpiles.org/
http://www.fao.org/ag/AGP/AGPP/Pesticid/Disposal/index_en.htm
http://www.sdpi.org/research_Programme/environment/Hazardous_Waste_Management.htm#2
http://ipen.ecn.cz/
http://www.envirohealthaction.org/toxics/pollution/
http://europa.eu.int/comm/environment/pops/index_en.htm
http://lnweb18.worldbank.org/ESSD/envext.nsf/50ByDocName/WhatArePOPs
http://www.deh.gov.au/industry/chemicals/international/pop.html

5.0   VENDOR CONTACTS

Full-Scale Technologies for Treatment of POPs

Anaerobic Bioremediation Using Blood Meal for Treatment of Toxaphene in Soil and Sediment
Mr. Harry L. Allen III, Ph.D.
EPA Environmental Response Team
MS-101, Building 18
2890 Woodbridge Avenue
Edison, NJ 08837
Telephone: (732) 321-6747
Fax: (732) 321-6724
Email: allen.harry@epa.gov

DARAMEND®
Dr. Alan G. Seech or Mr. David Raymond
Adventus Remediation Technologies, Inc.
1345 Fewster Drive
Mississauga, Ontario, Canada L4W 2A5
Telephone: (905) 273-5374, Extension 221
Mobile: (416) 917-0099
Fax: (905) 273-4367
Email: info@adventusremediation.com
Web site: http://www.adventusremediation.com

Gas-Phase Chemical Reduction (GPCR™)
Bennett Environmental Inc.
1540 Cornwall Road, Suite 208
Oakville, Ontario, Canada L6J 7W5
Telephone: (905) 339-1540
Fax: (905) 339-0016
Email: info@bennettenv.com
Web site: http://www.bennettenv.com

GeoMelt™
Mr. Kevin Finucane
AMEC Earth and Environmental, Inc.
309 Bradely Boulevard, Suite 115
Richland, WA 99352
Telephone: (509) 942-1292
Fax: (509) 942-1293
Email: kevin.finucane@amec.com
Web site: www.geomelt.com

In Situ Thermal Desorption (ISTD)
Mr. Ralph Baker
TerraTherm, Inc.
356 Broad Street
Fitchburg, MA 01420
Telephone: (978) 343-0300
Fax: (978) 343-2727
Email: rbaker@terratherm.com
Web site: www.terratherm.com

Mechanochemical Dehalogenation (MCD™)
Mr. Bryan Black
Environmental Decontamination Ltd.
P.O. Box 58-609
Greenmount
Aukland, New Zealand
Telephone: (649) 274-9862
Fax: (649) 274-7393
Email: bryan@manco.co.nz
Web site: http://edl.net.nz

Mr. Volker Birke
Tribochem
Georgstrasse 14
D-31515 Wunstdrof
Germany
Telephone: 49 5031 6 73 93
Fax: 49 5031 88 07
Email: birke@tribochem.com
Web site: www.tribochem.com

Xenorem™
Mr. Michael Klerkin
Technology Transfer Corporation
University of Delaware
Newark, DE 19716
Telephone: (302) 831-4230
Web site: http://www.udel.edu/

Pilot-Scale Technologies for Treatment of POPs

Base Catalyzed Decomposition
Mr. Terrence Lyons
EPA National Risk Management Research Laboratory
26 West Martin Luther King Drive
Cincinnati, OH 45268
Telephone: (513) 569-7589
Fax: (513) 569-7676

Mr. Charles Rogers
BCD Group Inc.
Cincinnati, OH
Telephone: (513) 385-4459

CerOx™
Mr. Matt van Steenwyk or Mr. Norvell Nelson
CerOx Corporation
2602 Airpark Drive
Santa Maria, CA 93455
Telephone: (805) 925-8111
Fax: (805) 925-8218
Email: mattvs@cerox.com/njnelson@cerox.com
Web site: www.cerox.com

Sonic Technology
Mr. Paul Austin
Sonic Environmental Solutions Inc.
1066 West Hastings Street, Suite 2100
Vancouver, British Columbia, Canada
V6E 3X2
Telephone: (604) 736-2552
Fax: (604) 736-2558
Email: paustin@sesi.ca
Web site: www.sonicenvironmental.com

Bench-Scale Technologies for Treatment of POPs

Self-Propagating High-Temperature Dehalogenation
Dr. Ing. Giacomo Cao Centro Studi Sulle Reazioni Autopropaganti Dipartimento di Ingegneria Chimica e Materiali Piazza d'armi 09123 Cagilari Italy
Telephone: 39-070-6755058
Fax: 39-070-6755057
Email: cao@visnu.dicm.unica.it

TDR-3R™
Mr. Edward Someus
Terra Humana Clean Technology Engineering Ltd.
1222 Budapest, Szechenyi 59
Hungary
Telephone: (36-20) 201 7557
Fax: (36-1) 424 0224
Email: edward@terrenum.net
Web site: http://www.terrenum.net

Full-Scale Technologies with Potential to Treat POPs

Plasma Arc Centrifugal Treatment
Mr. Leroy Leland
RETECH Systems
100 Henry Station Road
Ukiah, CA 95482
Telephone: (707) 467-1724
Fax: (707) 462-4103
Email: Leroy.b.leland@retechsystemsllc.com
Web site: www.retechsystemsllc.com

PLASCON™
Mr. Rex Williams or Mr. Martin Krynen
BCD Technologies Pty. Ltd.
Narangba, Queensland, Australia 4504
Telephone: 61 7 3203 3400
Fax: 61 7 3203 3450
E-mail: marius@gil.com.au

Plasma Converter System
Startech Environmental Corp.
15 Old Danbury Road
Wilton, CT. 06897-2525
Telephone: (203) 762-2499
Toll Free: (888) 807-9443
Fax: (203) 761-0839
Email: starmail@startech.net

Supercritical Water Oxidation
General Atomics
P.O. Box 85608
San Diego, CA 92186-5608
Telephone: (858) 455-3000
Fax: (858) 455-3621

Turbosystems Engineering Inc.
Telephone: (707) 529-7477
Fax: (707) 581-1749

6.0   REFERENCES

1.   Adventus Remediation Technologies, Inc. DARAMEND® project summaries. Online Address: http://www.adventusremediation.com.

2.   Allen, Harry L., and others. 2002. "Anaerobic Bioremediation of Toxaphene Contaminated Soil – A Practical Solution." 17th World Congress of Soil Science, Symposium No. 42, Paper No. 1509. Thailand. August 14 through 21.

3.   Allen, Harry L., EPA Environmental Response Team. 2005. Email to Younus Burhan, Tetra Tech EM Inc., Regarding EPA Comments on Draft Blood Meal Fact Sheet. January 25.

4.   American Chemical Society. 2000. "Herbicide and Pesticide Destruction." Symposium on Emerging Technologies: Waste Management in the 21st Century. San Francisco, California. May.

5.   Assembled Chemical Weapons Alternative. Program Summary. Online Address: http://www.pmacwa.army.mil/about/index.htm.

6.   Baker, Ralph, and Myron Kuhlman. 2002. "A Description of the Mechanisms of In-Situ Thermal Destruction (ISTD) Reactions." Second International Conference on Oxidation and Reduction Technologies for Soil and Groundwater (ORTs-2). Toronto, Ontario, Canada. November 17 through 21.

7.   Baker, Ralph, TerraTherm, Inc. 2004. Emails to Chitranjan Christian, Tetra Tech EM Inc. October 27 and November 8, 15, 24, and 29.

8.   Birke, Volker. 2002. "Reductive Dehalogenation of Recalcitrant Polyhalogenated Pollutants Using Ball Milling." Proceedings of the Third International Conference on Remediation of Chlorinated and Recalcitrant Compounds. Monterey, California.

9.   CerOx™ Corporation. 2005. Process Technology Overview. Online Address: http://www.cerox.com/systems_process.html.

10.   CMPS&F - Environment Australia. 1997. "Appropriate Technologies for the Treatment of Scheduled Wastes." Review Report Number 4. November. Online Address: http://www.oztoxics.org/research/3000_hcbweb/library/gov_fed/appteck/plasma.html#pact.

11.   Donnelly, Paul K., Hedge, Ramesh, S., and Fletcher, John S. (1994) "Growth of PCB-Degrading Bacteria on Compounds from Photosynthetic Plants." *Chemosphere*. Volume 28, Number 5. Pages 981-988.

12.   Eco Logic. 2002. "Contaminated Soil and Sediment Treatment Using the GPCRTM/TORBED® Combination." October. Online Address: http://www.torftech.com/start.htm.

13.   U.S. Environmental Protection Agency (EPA). 1996. "Cost and Performance Summary Report, Bioremediation at the Stauffer Management Company Superfund Site, Tampa, Florida." Office of Solid Waste and Emergency Response.

14.   EPA. 1994. "Eco Logic International Gas-Phase Chemical Reduction Process-The Thermal Desorption Unit." Superfund Innovative Technology Evaluation Program. EPA/540/AR-94/504. September. Online Address: http://www.epa.gov/ORD/SITE/reports/540ar94504/540ar94504.pdf

15.   EPA. 1996. Site Technology Capsule: "GRACE Bioremediation Technologies DARAMEND® Bioremediation Technology." Superfund Innovative Technology Evaluation Program. EPA/540/R-95/536.

16. EPA. 2000. "Cost and Performance Summary Report, Bioremediation at the Stauffer Management Company Superfund Site, Tampa, Florida." Office of Superfund Remediation and Technology Innovation. September.

17. EPA. 2000. "Potential Applicability of Assembled Chemical Weapons Assessment Technologies to RCRA Waste Streams and Contaminated Media." Office of Solid Waste and Emergency Response, Technology Innovation Office. August. EPA-R-00-004. Online Address: http://www.epa.gov/tio/download/remed/acwatechreport.pdf

18. EPA. 2002. "Persistent Organic Pollutants, A Global Issue, A Global Response." Office of International Affairs. EPA160-F-02-001.

19. EPA. 2004. "Cost and Performance Summary Report, The Legacy of the Navajo Vats Superfund Site, Arizona and New Mexico." Office of Superfund Remediation and Technology Innovation. October.

20. EPA. 2004. "Field Evaluation of TerraTherm In Situ Thermal Destruction (ISTD) Treatment of Hexachlorocyclopentadiene." Office of Research and Development, Superfund Innovative Technology Evaluation Program. EPA/540/R-05/007. July. Online Address: http://www.epa.gov/ORD/NRMRL/pubs/540r05007/540R05007.pdf

21. EPA. 2004. "In Situ Thermal Treatment of Chlorinated Solvents: Fundamentals and Field Applications." EPA 542-R-04-010. March.

22. EPA. 2004. T.H. Agricultural & Nutrition Company Site Information and Source Data. Online Address: http://www.epareachit.org.

23. EPA. 2005. Web Site on Persistent Organic Pollutants (POP). Office of Pesticide Programs. Information Downloaded on January 5. Online Address: http://www.epa.gov/oppfod01/international/pops.htm.

24. Finucane, Kevin, Geomelt. 2005. Emails to Ellen Rubin, EPA, Office of Superfund Remediation Technology Innovation, Regarding Geomelt ICV Process Description and Photo, and Results for WCS Rock Flats and WCS Commercial TSCA Cleanup. May 23.

25. General Atomics. 2005. Website of Advanced Process Systems Division. Online Address: http://demil.ga.com/.

26. GeoMelt® Technologies/AMEC Earth and Environmental, Inc., Richland, Washington. 2005. Web Page on GeoMelt® Technology Description. Online Address: http://www.geomelt.com/technologies/.

27. Gilbert, Eric S. and David E. Crowley. 1997. "Plant Compounds that Induce Polychlorinated Biphenyl Biodegradation by *Anthrobacter* sp. Strain B1B." *Applied and Environmental Microbiology*. Volume 63, Number 5. Pages 1933-1938.

28. Global Security. 2005. Weapons of Mass Destruction. Army Facilities. Web Page on Newport Chemical Depot (NECD), Newport, Indiana. Online Address: http://www.globalsecurity.org/wmd/facility/newport.htm.

29. Gray, N.C.C., AstraZeneca Group PLC. 2004. Telephone Conversation with Younus Burhan, Tetra Tech EM Inc. December 15.

30. Gray, N.C.C., P.R. Cline, A.L. Gray, B. Byod, G.P. Moser, H.A. Guiler, and D.J. Gannon. 2002. "Bioremediation of a Pesticide Formulation Plant." Proceedings of the Third International Conference on Remediation of Chlorinated and Recalcitrant Compounds. Monterey, California.

31. Hurt, K. 2005. "Successful Full Scale Phytoremediation of PCB and TPH Contaminated Soil," The Third International Phytotechnologies Conference, Atlanta, Georgia. April 19 - 22.

32. Ing, Giacomo Cao, Centro Studi sulle Reazioni Autopropaganti. 2004. Email to Younus Burhan, Tetra Tech EM Inc. December 12.

33. International HCH and Pesticides Association. NATO/CCMS Pilot Study Fellowship Report. Evaluation of Demonstrated and Emerging Remedial Action Technologies for the Treatment of Contaminated Land and Groundwater (Phase III). Online Address: http://www.ihpa.info/libraryNATO.htm.

34. Interstate Technology Regulatory Council. 2003. "Technology Overview Using Case Studies of Alternative Landfill Technologies and Associated Regulatory Topics."

35. Johnson, Lou, General Atomics. 2005. Telephone Conversation with Chitranjan Christian, Tetra Tech EM Inc. February 15.

36. Leigh, M., Fletcher, J., Nagle, D.P., Prouzova P., Mackova, M. and Macek, T. 2003. "Rhizoremediation of PCBs: Mechanistic and Field Investigations." Proceedings of the International Applied Phytotechnologies Conference. Chicago, Illinois.

37. Lyons, Terry, EPA National Risk Management Research Laboratory. 2005. Email to Younus Burhan, Tetra Tech EM Inc. January 19 and August 10.

38. Moklyachuk, L., Sorochinky, B. and Kulakow, P.A. 2005. "Phytotechnologies for Management of Radionucleide and Obsolete Pesticide Contaminated Soil in Ukraine," The Third International Phytotechnologies Conference, Atlanta, Georgia. April 19 - 22.

39. Nurzhanova, A., Kulakow, P., Rubin, E., Rakhimbaev, I., Sedlovsky, A., Zhambakin, K., Kalygin, S., Kalmykov, E. L. and Erickson. L. 2005. "Monitoring Plant Species Growth in Pesticide Contaminated Soil," The Third International Phytotechnologies Conference, Atlanta, Georgia. April 19 - 22.

40. Pan, Danny, Bennett Environmental Inc. 2005. Telephone Conversation with Younus Burhan, Tetra Tech EM Inc. January 18.

41. Phillips, T., G. Bell, D. Raymond, K. Shaw, and A. Seech. 2001. 6th International HCH and Pesticides Forum, Poznan, Poland. "DARAMEND® Technology for In Situ Bioremediation of Soil Containing Organochlorine Pesticides." March 20 - 22.

42. Raymond, David, Adventus Remediation Technologies, Inc. 2004. Telephone Conversation with Younus Burhan, Tetra Tech EM Inc. August 25.

43. Retech Systems LLC. 2005. Web Page on Plasma Arc Centrifugal Treatment (PACT) Systems. Online Address: http://www.retechsystemsllc.com/PACT%20webpagesC/index.htm.

44. Rogers, Charles, BCD Group Inc. 2005. Email correspondence with Younus Burhan, Tetra Tech EM Inc. December 9 and 13.

45. Sonic Environmental Solutions Inc.2005. Sonic Technology Treatment Process, Online Address: http://www.sesi.ca/.

46. SRL Plasma Pty. Ltd. 2005. Web Page on "PLASCON™ Electric-Arc Plasma Hazardous Waste Destruction Process." Online Address: http://www.srlplasma.com.au/srlpages/srlframe.html.

47. STARTECH Environmental Corp. 2005. Process Description. Web Page on Plasma Converter System. Online Address: http://www.startech.net/plasma.html.

48. Stegemeier, G.L., and Vinegar, H.J. 2001. "Thermal Conduction Heating for In-Situ Thermal Desorption of Soils," Chapter 4.6, pages 1-37. Chang H. Oh (ed.), Hazardous and Radioactive Waste Treatment Technologies Handbook, CRC Press, Boca Raton, FL.

49. Stockholm Convention on Persistent Organic Pollutants (POPS). 2005. Web Site for the Stockholm Convention. Online Address: http://www.pops.int/

50. TerraTherm Environmental Services. 1999. "Naval Facility Centerville Beach, Technology Demonstration Report, In-Situ Thermal Desorption." November.

51. TerraTherm, Inc. 2005. Web Page on Technology Process Description (ISTD). Online Address: http://www.terratherm.com/default.htm.

52. Thermal Desorption Technology Group LLC of North America. 2005. Website of Terra Humana Clean Technology Engineering Ltd. Online Address: http://www.terrenum.net/index.htm.

53. Thermal Desorption Technology Group, Terra Humana Clean Technology Engineering Ltd. 2004. "Summary Report of the TDT-3R Treatment - Latest Five Years – Projects 2000-2004." December 10.

54. Thiess Services NSW. 2004. "Proof of Performance Report, FCC Remediation, Mapua, New Zealand." June.

55. Turbosystems Engineering Inc. 2005. Web Page on Supercritical Water Oxidation Technology. Online Address: http://www.turbosynthesis.com/summitresearch/sumhome.htm.

56. U.S. Department of Health and Human Services. 2005. Toxicological Profile Information Sheet. Agency for Toxic Substances and Disease Registry (ATSDR). Online Address: http://www.atsdr.cdc.gov/toxpro2.html.

57. United Nations Environment Programme (UNEP). 2003. Science and Technology Advisory Panel (STAP) of the Global Environmental Facility. "Report of the STAP/GEF POPs Workshop on Non-Combustion Technologies for the Destruction of POPs Stockpiles." October. Online Address: http://www.basel.int/techmatters/review_pop_feb04.pdf

58. UNEP. 2004. STAP of the GEF. "Review of Emerging, Innovative Technologies for the Destruction and Decontamination of POPs and the Identification of Promising Technologies for Use in Developing Countries." GF/8000-02-02-2205. January. Online Address: http://www.unep.org/stapgef/home/index.htm.

59. UNEP. 2005. "Technical Guidelines for the Environmentally Sound Management of Persistent Organic Pollutant Wastes." UNEP/POPS/COP.1/11. May. Online Address: http://www.pops.int/documents/meetings/cop_1/meetingdocs/en/cop1_11/COP_1_11.pdf.

60. UNEP. 2005. Web Site of the Secretariat of the Basel Convention. Online Address: http://www.basel.int/

61. White, J. C., Mattina, M. I., Eitzer, B. D., Isleyen, M., Parrish, Z. D. and Gent, M. P.N. 2005. "Enhancing the Uptake of Weathered Persistent Organic Pollutants by Cucurbita pepo," The Third International Phytotechnologies Conference, Atlanta, Georgia. April 19 - 22.

62. Zeeb, B., Whitfield, M. and Reimer, K. J. 2005. "In situ Phytoextraction of PCBs from Soil: Field Study," The Third International Phytotechnologies Conference, Atlanta, Georgia. April 19 - 22.

APPENDIX A

Anaerobic Bioremediation Using Blood Meal for the Treatment of Toxaphene in Soil and Sediment

POPS-WASTES APPLICABILITY (REFS. 1 AND 5):

Anaerobic Bioremediation Using Blood Meal was able to rapidly degrade toxaphene in soil to achieve cleanup goals in bench- and pilot-scale tests. Bench-scale tests have indicated that the technology is also effective in treating dichlorodiphenyltrichloroethane (DDT). Full-scale implementations have successfully treated several toxaphene-contaminated sites. The quantity of soil treated at these sites ranged from 250 to 8,000 cubic yards. This technology does not typically achieve greater than 90 percent contaminant reduction.

POPs Treated:	Toxaphene and DDT
Other Contaminants Treated:	None
Application:	Ex-situ

TECHNOLOGY DESCRIPTION (REFS. 1 AND 5):

OVERVIEW

This technology uses biostimulation to accelerate the degradation of toxaphene in soil or sediment. It involves the addition of biological amendments, including blood meal (nutrient) and phosphates (pH buffer), to stimulate native anaerobic microorganisms. Blood meal is a black powdery fertilizer made from animal blood. The typical dosage of blood meal and sodium phosphate is one percent by weight of contaminated soil. This is sometimes augmented with one percent by weight of starch to rapidly establish anaerobic conditions. The standard recipe uses monobasic and dibasic phosphate salts in equal proportions (monobasic:dibasic - 1:1) to maintain soil pH around 6.7. The low phosphate/starch recipe uses three times more dibasic than monobasic phosphates (monobasic:dibasic – 1:3) and maintains soil pH around 7.8.

The soil to be treated is mixed with amendments and water. Mixing methods including blending in a dump truck, mechanical mixing in a pit, and mixing in a pug mill have been used to produce homogeneous soil-amendment mixtures. The mixture is transferred to a cell with a plastic liner, and excess water is added to provide up to a foot of cover above the settled solids. The water provides a barrier that minimizes the transfer of atmospheric oxygen to microorganisms in the slurry, which helps maintain anaerobic conditions. The lined cell is covered with a plastic sheet to isolate the cell from the environment, and the slurry is incubated for several months. The slurry may be sampled periodically to measure treatment progress. Once treatment goals have been met, the cell is drained. The slurry is usually left in place, but it may be dried and used as fill material on site. The slurry also serve as a source of acclimated microorganisms for use at another toxaphene-contaminated site.

Anaerobic degradation of toxaphene usually results in the production of intermediates such as less chlorinated congeners of toxaphene. Further degradation of intermediates results in the production of carbon dioxide, methane, water, inorganic chlorides, and cell mass.

STATUS AND AVAILABILITY (REFS. 2 AND 6):

The technology has been implemented at full scale to treat toxaphene-contaminated sites. Four such sites are:

(1) The Laahty Family Dip Vat (LDV) site (253 cubic yards in one cell)
(2) The Henry O Dip Vat (HDV) site (660 cubic yards in two cells)
(3) The Gila River Indian Community (GRIC 1) site (3,500 cubic yards in four cells)
(4) The GRIC 2 site (8,000 cubic yards in five cells)

EPA's Environmental Response Team (ERT) is the developer of the technology. The technology is unlicensed and is available through the ERT. The biological amendments (blood meal, and monobasic and dibasic phosphates) are inexpensive and commercially available.

Design (Refs 1, 5):
Factors that need to be considered when designing an anaerobic bioremediation process using blood meal include:
- The presence of active toxaphene-degrading bacteria
- Soil characteristics
- Volume of soil to be treated
- Concentration of toxaphene in contaminated soil
- Cleanup goal
- Availability of space on site for the construction of treatment cells
- Odor mitigation requirements as determined by surrounding land use and the proximity of residences
- Need for agreements with landowners and community leaders
- Climate
- Security issues
- Availability of water

THROUGHPUT (REFS. 1 AND 5):
Throughput of a technology that does not operate like a batch processing plant is hard to define. Remediation involves a series of steps including construction, mix preparation, and treatment. Treatment is usually the slowest step. Factors that can influence treatment time include, the type of microbial communities present, amendment dosage, contaminant concentration, treatment goals, and the presence of inhibitors (such as very cold environments). In general, treatment time can vary from five weeks to two years.

WASTES/RESIDUALS (REFS 2, 3 AND 6):
Products of toxaphene degradation include lower-chlorinated chlorobornane congeners, chloride ions, cell mass, carbon dioxide, and methane. Chlorobornane congeners have been shown to degrade completely during treatment. However, treated soil can contain low concentrations (below cleanup goals) of unutilized toxaphene and lower-chlorinated chlorobornane congeners.

Gaseous wastes produced can include methane and hydrogen sulfide. Therefore, odor concerns should be considered. If treatment cells are not left in place at the end of remediation, solid wastes can include debris from the demolition of treatment cells and associated temporary facilities. Debris potentially contaminated with toxaphene will require testing to determine its hazardous nature in compliance with local, State, and Federal requirements prior to disposal.

MAINTENANCE (REFS. 2 AND 6):
- Periodic addition of water to treatment cells to maintain water level
- Maintaining treatment cells to prevent leaks
- Maintaining cover integrity
- Monitoring for gas buildup
- Monitoring for fugitive odors
- Soil sampling to monitor remedial progress

LIMITATIONS (REFS. 2 AND 6):
- The anaerobic process is affected by temperature. Spring and summer are the best periods for operation. This technology cannot be used in extremely cold climates.
- This technology requires a bench scale test to determine applicability at a given site, and to estimate treatment duration.
- At a minimum, five weeks are required for treatment.

- This technology typically does not achieve greater than 90 percent contaminant destruction.
- Blood meal accelerates the rate of reductive dechlorination of toxaphene, but does not affect the extent of dechlorination.
- Unfavorable soil chemistry can inhibit the process. Unfavorable soil chemistry may result from the presence of bioavailable heavy metals including mercury, arsenic, and chromium; solvents; and pesticides (including toxaphene).
- Level C personal protective equipment is required when working with blood meal.

FULL-SCALE TREATMENT EXAMPLES (REFS. 1, 2, 5 AND 6):
Anaerobic bioremediation using blood meal and phosphate amendments has been implemented at a full scale at twenty two (22) Dip Vat sites in the Navajo Nation. Other sites where this technology has been applied at a full scale to remediate toxaphene-contaminated soil include:

(1) The Ojo Caliente Dip Vat site
(2) The Laahty Family Dip Vat site
(3) The Henry O Dip Vat site
(4) The Acoma Reservation at Sky City
(5) The Gila River Indian Community (GRIC 1) crop duster site
(6) The GRIC 2 crop duster site

The resources used for this fact sheet contain performance data on nine applications of this technology. Performance data for each of these sites is presented in Table 1 at the end of this fact sheet. Three of these sites are discussed below in greater detail. The unit cost of implementation at these sites in USD ranged from $98 to $296 per cubic yard.

Laahty Family Dip Vat (LDV) site

The LDV site is located in The Zuni Nation, New Mexico. Soil at the site was contaminated with toxaphene at an average concentration of 29 milligrams per kilogram (mg/kg). A total of 253 cubic yards (cy) of soil was excavated and stockpiled on site. A cell with dimensions, 73 feet (ft) by 30 ft by 4 ft (deep) was constructed and lined with a plastic liner. Contaminated soil was placed in a concrete mixer and mixed with biological amendments and water. Blood meal and monobasic phosphate were added, each at a dosage rate of 10 grams per kilogram (g/kg) of contaminated soil. Dibasic phosphate salts were also added at a dosage rate of 3.3 g/kg soil. The nutrient-amended soil slurry was then placed in the lined cell. Water was added to provide one foot of cover above the solids in the cell. The cell was then covered with a plastic sheet and incubated. Samples were collected periodically to monitor progress. The toxaphene concentration decreased in the anaerobic cell from an initial concentration of 29 mg/kg to 4 mg/kg in 31 days. This corresponded to an overall reduction of 86 percent. The post-treatment concentrations were below the 17 mg/kg action level established for the site. In 2004, the total cost of treatment in USD was $75,000. Consequently, the unit cost of treatment at this site was $296 per cubic yard.

Henry O Dip Vat (HDV) Site

The HDV site is located in The Zuni Nation, New Mexico. Approximately 660 cy of soil at this site was contaminated with toxaphene at an average concentration of 23 mg/kg. Two cells were constructed for soil treatment:

- The north cell (Cell 1) was 75 ft by 35 ft by 5 ft (deep).
- The south cell (Cell 2) was 65 ft by 30 ft by 5 ft (deep).

Both cells were lined with plastic liners. Blood meal and sodium phosphate were added to contaminated soil and placed in a mixing pit using a backhoe. The dosage rate of blood meal was 5 g/kg of contaminated soil, while that of monobasic phosphate was 10 g/kg of contaminated soil.

Dibasic phosphate salts were also added at a dosage rate of 3.3 g/kg. Water was added to the soil in the mixing pit, and the resulting soil slurry was extensively mixed. Once mixed, the soil slurry was transferred to anaerobic cells 1 and 2. Water was added to provide one foot of additional cover above the solids in each cell. Each cell was then covered with a plastic sheet and incubated for 61 to 76 days. Samples were collected on day 1 and day 61 from Cell 1 and on day 1 and 76 from Cell 2. Analysis of the samples indicated that the average toxaphene concentration was reduced from 23 mg/kg to 8 mg/kg. This corresponds to a percent removal of approximately 67 percent removal in 68 days. The post-treatment concentrations were below the 17 mg/kg action level established for the site. In 2004, the total cost of treatment in USD was $65,000. Consequently, the unit cost of treatment at this site was$98 per cubic yard.

Gila River Indian Community Site

The Gila River Indian Community (GRIC) site is located in Chandler, Arizona. Approximately 3,500 cy of toxaphene-contaminated soil required treatment at this site. Four lined cells were constructed with dimensions of 178 ft by 43 ft by 7 ft (deep). This dosage rate was lower than for other sites to reduce costs. The dosage rate of blood meal, sodium phosphate, and dibasic phosphates was 5 g/kg of contaminated soil. Blood meal and phosphates were first mixed in a pit, and then blended with contaminated soil using a pug mill (100-300 cy/hr throughput). The mixture was then transferred to cells filled with water to 25 percent capacity. Additional water was then added to the cells to provide one foot of cover above the solids. Each cell was then covered with a plastic sheet. Samples were collected from the cells after initial setup and at the end of 3 months, 6 months, and 9 months. The removal of toxaphene in GRIC site soil took longer than usual due to the reduced amendment dosage rates. The average toxaphene concentration at the end of 180 days ranged between 4 mg/kg and 5 mg/kg demonstrating 83 to 88 percent toxaphene removal. The samples collected at day 272 showed residual levels of 2 to 4 mg/kg corresponding to a percent removal between 87 and 98 percent. The post-treatment concentrations were below the 17 mg/kg action level established for the site. In 2004, the total cost of treatment in USD was $793,000. Consequently, the unit cost of treatment at this site was $226 per cubic yard.

Table 1
Performance Data for Anaerobic Bioremediation of Toxaphene Using Blood Meal at Selected Sites

Site Name	Untreated Concentration (mg/kg)	Treated Concentration (mg/kg)	Period (Days)	Percent Reduction	Volume Treated (cy)
Navajo Vats Chapter					
Nazlini	291	71	108	76	NA
Whippoorwill	40	17	110	58	NA
Blue Canyon Road	100	17	106	83	NA
Jeddito Island	22	3	76	77	NA
Poverty Tank	33	8	345	76	NA
Ojo Caliente	14	4	14	71	200
Laahty Family Dip Vat	29	4	31	86	253
Henry O Dip Vat	23	8	68	67	660
Gila River Indian Community					
Gila River Indian Community (Cell 1)	59	4	272	94	
Gila River Indian Community (Cell 2)	31	4	272	87	
Gila River Indian Community (Cell 3)	29	2	272	94	
Gila River Indian Community (Cell 4)	211	3	272	98	3,500

Note:
mg/kg: Milligrams per kilogram
NA: Not available
Source: Refs. 1, 2 and 6

U.S. EPA CONTACT:	LAAHTY FAMILY AND HENRY O DIP VAT SITES:	Gila River Indian Community CONTACT:
U.S. EPA Environmental Response Team Harry L. Allen III, Ph.D. Phone: (732) 321-6747 Email: allen.harry@epa.gov	Bureau of Indian Affairs Southwest Region Zuni Nation Phone: (505) 563-3106	GRIC Department of Environmental Quality Hazardous Waste Program Manager Dan Marsin Email: hazmat@gilnet.net Phone: (520) 562-2234

PATENT NOTICE:
This technology has not been patented.

REFERENCES:

1. Allen L., Harry and others. 2002. Anaerobic bioremediation of toxaphene-contaminated soil – a practical solution. 17th WCCS, Symposium No. 42, Paper No. 1509, Thailand. August 14 – 21.

2. Allen L., Harry, EPA Environmental Response Team. 2005. Email to Younus Burhan, Tetra Tech EM Inc., Regarding Comments from Harry L. Allen on Draft (January 5, 2005) Blood Meal Fact Sheet. January 25.

3. Allen L., Harry, EPA Environmental Response Team. 2005. Memo to Ellen Rubin, EPA Office of Superfund Remediation and Technology Innovation. Response to Questions on Toxaphene Fact Sheet. February 24.

4. U.S. Environmental Protection Agency (EPA). Office of Superfund Remediation and Technology Innovation. 2004. Cost and Performance Summary Report. The Legacy of the Navajo Vats Superfund Site, Arizona and New Mexico. October.

5. EPA. 2000. Fact Sheet - Gila River Indian Community Toxaphene Site. October.

6. Rubin, Ellen, EPA Environmental Response Team. 2005. Email to Younus Burhan, Tetra Tech EM Inc., Regarding Comments from Dr. T. Ferrell Miller on Draft (January 5, 2005) Blood Meal Fact Sheet. February 7.

APPENDIX B

Bioremediation Using DARAMEND® for Treatment of POPs in Soils and Sediments

POPs - WASTES APPLICABILITY (REFS. 1, 6, AND 10):
DARAMEND® is a bioremediation technology that has been used to treat soils and sediments containing low concentrations of pesticides such as toxaphene and DDT as well as other contaminants.

POPs Treated:	Toxaphene and DDT
Other Contaminants Treated:	DDD, DDE, RDX, HMX, DNT, and TNT

TECHNOLOGY DESCRIPTION (REFS. 4, 5 AND 10):

OVERVIEW

DARAMEND® is an amendment-enhanced bioremediation technology for the treatment of POPs that involves the creation of sequential anoxic and oxic conditions. The treatment process involves the following:

1. Addition of solid phase DARAMEND® organic soil amendment of specific particle size distribution and nutrient profile, zero valent iron, and water to produce anoxic conditions.
2. Periodic tilling of the soil to promote oxic conditions.
3. Repetition of the anoxic-oxic cycle until the desired cleanup goals are achieved.

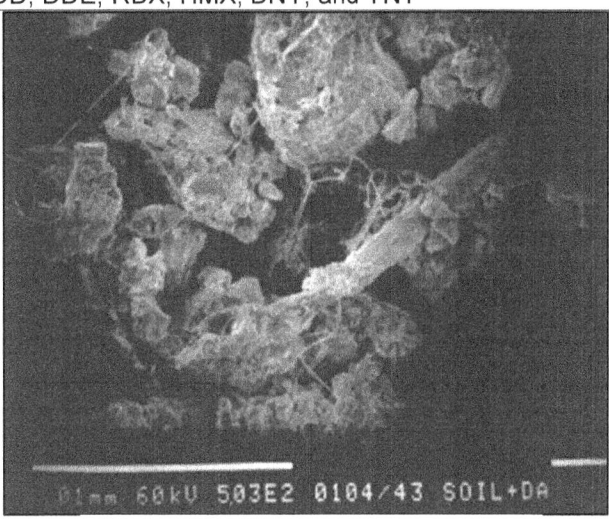
DARAMEND® particle colonization as viewed through an electron-microscope

The addition of DARAMEND® organic amendment, zero valent iron, and water stimulates the biological depletion of oxygen generating strong reducing (anoxic) conditions within the soil matrix. The diffusion of replacement oxygen into the soil matrix is prevented by near saturation of the soil pores with water. The depletion of oxygen creates a very low redox potential, which promotes dechlorination of organochlorine compounds. A cover may be used to control the moisture content, increase the temperature of the soil matrix and eliminate run-on/run off. The soil matrix consisting of contaminated soil and the amendments is left undisturbed for the duration of the anoxic phase of treatment cycle (typically 1- 2 weeks).

In the oxic phase of each cycle, periodic tilling of the soil increases diffusion of oxygen to microsites and distribution of irrigation water in the soil. The dechlorination products formed during the anoxic degradation process are subsequently removed trough aerobic (oxic) biodegradation processes, initiated by the passive air drying and tilling of the soil to promote aerobic conditions.

Addition of DARAMEND® and the anoxic-oxic cycle continues until the desired cleanup goals are achieved. The frequency of irrigation is determined by weekly monitoring of soil moisture conditions. Soil moisture is maintained within a specific range below its water holding capacity. Maintenance of soil moisture content within a specified range facilitates rapid growth of an active microbial population and prevents the generation of leachate. The amount of DARAMEND® added in the second and subsequent treatment cycles is generally less than the amount added during the first cycle.

DARAMEND® technology can be implemented using land farming practices either ex situ or in situ. In both cases, the treatment layer is 2 feet (ft) deep, the typical depth reached by tilling equipment. However, the technology can be implementation in 2-ft sequential lifts. In the ex situ process, the contaminated soil is excavated and sometimes mechanically screened in order to remove debris that may interfere with the distribution of the organic amendment. The screened soil is transported to the treatment unit, which is typically an earthen or concrete cell lined with a high-density polyethylene liner. In situ, the soil may be screened to a depth of 2-ft using equipment such as subsurface combs and agricultural rock pickers.

STATUS AND AVAILABILITY (REF. 1):
DARAMEND® is a proprietary technology and is available only through one vendor - Adventus Remediation Technologies (ART), Mississauga, Ontario, Canada. In the U.S., the technology is provided by ART's sister company, Adventus Americas Inc., Bloomingdale, IL. The technology has been used for the treatment of POPs (toxaphene and DDT) since 2001. Table 1 lists performance data for DARAMEND® technology application at selected sites. Through 2005, DARAMEND® has been implemented at two POPs contaminated sites.

Site Name	Scale	Quantity Treated (tons)	No. of treatment cycles	Duration of each cycle	Cost per ton*	Contaminant	Untreated Concen-tration (mg/kg)	Treated Concen-tration (mg/kg)
Table 1: Performance Data of DARAMEND at Selected Sites						Performance		
POPs Contaminated Sites								
T.H. Agricultural & Nutrition (THAN) Superfund Site, Montgomery, Alabama	Full	4,500	15	10 days	$55	Toxaphene DDT DDE DDD	See Table 2 for performance data	
W.R. Grace, Charleston, South Carolina	Pilot	250	8	1 month	$95	Toxaphene	239	5.1
						DDT	89.7	16.5
Non-POPs Contaminated Sites								
Naval Weapons Station, Yorktown, Virginia	Full	4,800	12	7-10 days	$90	TNT	15,359	14
						RDX	1,090	1.6
						DNT	1,002	13
Iowa Army Ammunition Plant, Burlington, Iowa	Full	8,000	5	7-10 days	$150	RDX	1,530,	16.2
						HMX	1,112,	84.5
						TNT	95.8	8
Confidential Site, Northwest U.S.A. (applied in multiple 2-ft lifts)	Full	6,000	Aerobic treatment	N/A	$37	PCP	359	8
						PCP	760	31

Source: Ref. 1
* Treatment costs are as reported by vendor. The vendor did not specify what was included in this cost.

DESIGN (REF. 5):
The major design factor for the implementation of this technology is the amount and type of soil amendments required for bioremediation. This is dependent on site conditions and the physical (textural variation, percent organic matter, and moisture content) and chemical (soil pH, macro and micronutrients, metals, concentration and nature of contaminants of concern) properties of the target soil. The duration of the treatment cycle is based on soil chemistry, concentration of contaminants of concern and soil temperature. The number of treatment cycles is based on the required cleanup levels of the contaminant.

THROUGHPUT (REF. 4):
For ex situ treatment, the amount of POPs contaminated soil/sediment that can be treated is dependent on the available surface area to spread contaminated soil. The technology can also be applied ex-situ in windrows. For in-situ application, the tillage equipment limits the depth (2-ft) to which the soil can be remediated. However, the technology can be used in-situ at depth greater than 2-ft using alternative soil mixing equipment or injection techniques.

WASTES/RESIDUALS (REF. 4):
The primary wastes generated are debris, stone, and construction material that are removed in the pretreatment process. No leachate is generated if a treatment area cover is used. If no cover is used, precipitation in the treatment area may generate leachate or storm water run-off.

Sampling and monitoring activities of the treatment pile will generate personal protective equipment (PPE) and contaminated water from decontamination activities.

MAINTENANCE:
Implementation of the DARAMEND® technology to treat POPs requires limited maintenance such as the upkeep of tilling, soil moisture control, and other industrial equipment. Because the specific amendments and application rate of DARAMEND® are site and soil-specific, the ongoing maintenance will vary by site and type of soil treated.

LIMITATIONS (REFS. 4 AND 9):
DARAMEND® technology may become technically or economically infeasible when treating soils with excessively high contaminant concentration. The technology has not been used for the treatment of other POPs such as PCBs, dioxins, or furans. ART, the developer of the technology, indicated that it has been only marginally successful in bench scale treatment of PCB-contaminated soil. Bench scale or pilot scale studies are typically conducted before field application of this technology; the type and amount of soil amendments required are then based on the results of these studies.

In situ application of this technology using tilling equipment is limited to a depth of 2-ft. However, the technology can be used in situ at depths greater than 2-ft using alternative soil mixing equipment or injection techniques. This technology requires that the treatment area be free of surface and subsurface obstructions that would interfere with the soil tilling. Ex situ application of this technology requires a large surface area to treat large quantities of the contaminated soil. Implementation of this technology in 2-ft sequential lifts would increase the total time required to treat the contaminated soil. The technology can also be applied ex situ in windrows.

Application of this technology requires a source of water (either city, surface, or subsurface).

This technology cannot be applied to sites that are prone to seasonal flooding or have a water table that fluctuates to within 3-ft of the site surface. These conditions make it difficult to maintain the appropriate range of soil moisture required for effective bioremediation, and may redistribute contamination across the site.

Volatile organic compound emissions may increase during soil tilling. Other factors that could interfere with the process would be large amounts of debris in the soil, which would interfere with the incorporation of organic amendments and reduce the effectiveness of tilling. Presence of other toxic compounds (heavy metals) may be detrimental to soil microbes. Soils with high humic content may slow down the cleanup through increased organic adsorption and oxygen demand.

FULL-SCALE TREATMENT EXAMPLES (REF. 3):
Bioremediation of pesticides-impacted soil/sediment, T.H. Agriculture and Nutrition (THAN) Superfund Site, Montgomery, Alabama.

The THAN site is located on the west side of Montgomery, Alabama, about 2 miles south of the Alabama River. The site is approximately 16 acres in area. Previous site operations involved the formulation, packing and distribution of pesticides, herbicides, and other industrial/waste treatment chemicals. The site was listed on the National Priorities List (NPL) on August 30, 1990. In 1991, EPA entered into a consent agreement with Elf Atochem North America Inc., the Potentially Responsible Party (PRP) for the site, to conduct a remedial investigation/feasibility study for the site. The final Record of Decision (ROD) for the site was signed on September 28, 1998, and bioremediation was selected as the remedy for treating the contaminated soils and sediments. DARAMEND® was selected as the bioremediation technology.

The contaminated soil and excavated sediments (approximately 4,500 tons) were treated using anaerobic/aerobic bioremediation cycle using DARAMEND®. Implementation of the technology involved the following steps:

1. DARAMEND® amendment and powdered iron application and incorporation
2. Determination of water holding capacity (first cycle only)
3. Determination of treatment matrix moisture content
4. Irrigation
5. Measurement of soil redox potential
6. Soil allowed to stand undisturbed for anoxic phase (approximately 7 days)
7. Soil tilled daily to generate oxic condition (approximately 4 days)
8. Steps 1, and 3 to 7 were repeated for each subsequent cycle. Fifteen treatment cycles were implemented in some treatment areas on site.

Two agricultural tractors (Model: Massey-Ferguson 394 H) mounted with deep rotary tillers were used for amendment application and tilling the treatment area. The target soil moisture content at the beginning of each cycle was approximately 33% (dry wt. basis) or 90% of the soil's water holding capacity. The optimal pH range (6.6 to 8.5) of the treatment area was maintained by adding hydrated lime at a rate of 1,000 mg/kg during the oxic phase of the third, sixth, and twelfth cycle. Following the application of each treatment cycle, samples were collected from the treatment area. The treatment area was divided into 12 sampling zones and one composite sample (composite of four grab samples) was collected from each zone. The samples were collected from the full 2-ft soil profile of treatment area. Fifteen treatment cycles were applied to some areas of the site. Table 2 lists the initial and final concentration of the samples collected from these 12 zones.

Based on the final sampling event DARAMEND® reduced the concentration of all the contaminants of concern to less than the specified performance standards. The average treatment cost in USD at the THAN site was $55 per ton. The vendor did not specify what was included in this cost.

Table 2: DARAMEND® performance at the THAN Site							
	Toxaphene (29 mg/kg)[1]		DDT (94 mg/kg)[1]		DDD (94 mg/kg)[1]		DDE (133 mg/kg)[1]	
Sampling Zone	Initial[2] Conc. (mg/kg)	Final[3] Conc. (mg/kg)	Initial[2] Conc. (mg/kg)	Final[3] Conc. (mg/kg)	Initial[2] Conc. (mg/kg)	Final[3] Conc. (mg/kg)	Initial[2] Conc. (mg/kg)	Final[3] Conc. (mg/kg)
1	77	< 20	126	10.2	52	26.4	33	6
2	260	< 21	227	15	133	73	35.3	8.4
3	340	< 21	33.2	4.5	500	89	49	7.8
4	45	< 21	55.1	14.7	34	37	15.8	7.2
5	230	< 21	216	16.1	93	53	22.4	6.8
6	90	< 21	13.3	2.2	130	59	17	5.7
7	100	< 20	151	15.3	85	38	25.2	6.3
8	13	< 20	9.1	5.2	44	24.3	6.9	2.8
9	330	< 21	45	5.7	312	85	28.2	7.2
10	48	< 20	44.4	5.7	146	25.5	20.1	4.2
11	20	< 20	12.6	2.9	46	25.1	6.9	3.0
12	720	< 21	78	6.3	590	87	59.6	8.6

Notes:
1. Performance Standard as specified in the Record of Decision, Summary of Remedial Alternatives Selection, THAN Site.
2. Initial concentration reported from samples collected by responsible party.
3. Final concentration reported from splits samples collected by EPA.

U.S. EPA RPM FOR THAN SITE:
Brian Farrier
EPA Region 4
Telephone: 404-562-8952
Fax: 404-562-8955
Email: farrier.brain@epa.gov

VENDOR CONTACT DETAILS:
David Raymond
Adventus Remediation Technologies, Inc.
1345 Fewster Drive
Mississauga, Ontario L4W 2A5
Telephone: 905-273-5374, Extension 224
Mobile: 416-818-0328
Fax: 905-273-4367
Email: info@adventusremediation.com
Web Site: http://www.adventusremediation.com

PATENT NOTICE:
DARAMEND® is a patented technology with U.S. Patent No. 5,618,427.

REFERENCES:

1. Adventus Remediation Technologies, Inc. DARAMEND project summaries. Online Address: http://www.adventusremediation.com.

2. Adventus Remediation Technologies, Inc. March 2002. Draft Final Report, Ex-Situ DARAMEND Bioremediation of Soil Containing Organic Explosive Compounds, Iowa Army Ammunition Plant, Middletown, Iowa.

3. Adventus Remediation Technologies, Inc. November 2003. Final Report, Bioremediation of Soil and Sediment Containing Chlorinated Organic Pesticides, THAN Superfund Site, Montgomery, Alabama.

4. EPA. 1996. Site Technology Capsule, GRACE Bioremediation Technologies DARAMEND® Bioremediation technology. Superfund Innovative Technology Evaluation. EPA/540/R-95/536.

5. EPA. 1997. Site Technology Capsule, GRACE Bioremediation Technologies DARAMEND® Bioremediation technology. Superfund Innovative Technology Evaluation. EPA/540/R-95/536a.

6. EPA. 2002. Technology News and Trends, Full-Scale Bioremediation of Organic Explosive contaminated soil. EPA 542-N-02-003. July.

7. EPA. 2004. TH Agricultural & Nutrition Company Site Information and Source Data. Online Address: http://www.epareachit.org.

8. EPA. 2004. TH Agricultural & Nutrition Company Site, RODS Abstract information, Superfund Information Systems. Online Address: http://www.epa.gov/superfund.

9. Farrier, Brian, EPA Region 4. 2004. Telephone Conversation with Younus Burhan, Tetra Tech EM Inc. August 31 and October 19.

10. Phillips, T., Bell, G., Raymond, D., Shaw, K., and Seech, A. 2001. "DARAMEND® technology for in situ bioremediation of soil containing organochlorine pesticides."

11. Raymond, David, Adventus Remediation Technologies, Inc. 2004. Telephone Conversation with Younus Burhan, Tetra Tech EM Inc. August 25.

APPENDIX C

In Situ Thermal Desorption for Treatment of POPs in Soils and Sediments

POPs-Wastes Applicability (Refs. 4 and 16):
ISTD is a thermally enhanced in-situ treatment technology that uses conductive heating elements to directly transfer heat to environmental media. ISTD can heat soil or sediment in situ to average temperatures of 1,000 degrees Fahrenheit (°F), and as a result has been used to treat compounds with relatively high boiling points. Some of these include semivolatile organic contaminants (SVOCs) such as polychlorinated biphenyls (PCBs), polycyclic aromatic hydrocarbons (PAHs), pesticides, and herbicides. Pilot- and full-scale applications have been performed where ISTD has been used to remove PCBs, and where dioxins and furans were trace contaminants. TerraTherm is the sole vendor for ISTD. According to TerraTherm, laboratory-scale work and extrapolation techniques have suggested the potential applicability of ISTD to POPs other than PCBs, dioxins, and furans (including aldrin, dieldrin, endrin, chlordane, heptachlor, DDT, mirex, hexachlorobenzene, and toxaphene); however, these contaminants have not yet been treated using ISTD on a full- or pilot-scale basis. ISTD has been used to treat contaminants in most hydrogeologic settings, including beneath structures.

POPs Treated:	PCBs, dioxins, and furans, aldrin, chlordane, dieldrin, and endrin
Other Contaminants Treated:	Hexachlorocyclopentadiene, isodrin, VOCs, SVOCs, oils, creosotes, coal tar PAHs, gasoline and diesel range organics, and MTBE

Technology Description (Refs. 2, 4, 13 and 16):
Overview
ISTD involves simultaneous application of heat and vacuum to subsurface soils. There are three basic elements in an ISTD process: (1) application of heat to contaminated media; (2) collection of desorbed contaminants through vapor extraction; and (3) treatment of collected vapors. Figure 1 presents a typical ISTD system.

ISTD has been used at full scale to treat PCBs, PAHs, dioxins, and chlorinated volatile organic compounds (CVOC). At the temperatures achieved by the ISTD process, volatiles metals such as mercury may also be recovered.

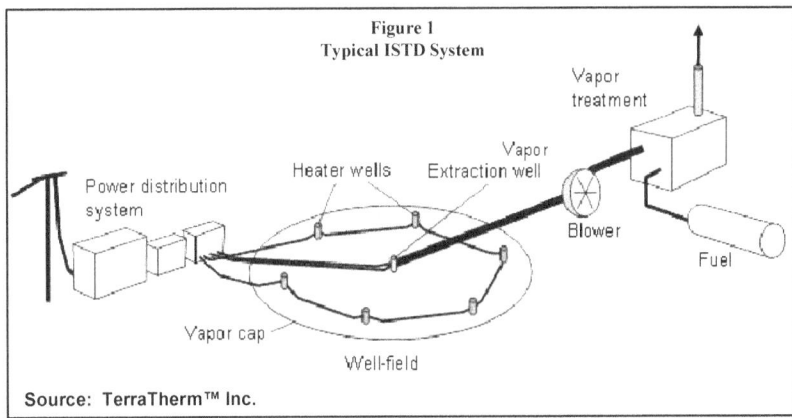

Figure 1
Typical ISTD System

Source: TerraTherm™ Inc.

In-Situ Heating

ISTD uses surface or buried electrically powered heaters to heat contaminated media. The most common setup uses a vertical array of heaters placed inside wells drilled into the remediation zone. A less common setup uses the same type of heaters installed horizontally on the surface of the contaminated zone. This method of heating (often called blanket heating) is typically used when contamination is shallow (usually 1 to 3 feet below ground surface (bgs)). Figure 2 illustrates the two different methods of heating.

ISTD heaters can attain temperatures as high as 1,600 degrees °F, and can produce average media temperatures exceeding 1,000 °F. Heat originates from a heating element and is transferred to the subsurface largely via thermal conduction and radiant heat transport, which dominates near the heat sources. There is also a contribution through convective heat transfer that occurs during the formation of steam from pore water present in the soil or sediment.

The thermal conductivity values of a wide range of soil types (e.g., clay, silt, sand, gravel) vary only by a factor of approximately four. Therefore, the rate of heat transfer from the linear heaters to the surrounding media is radially uniform. When heating commences, the temperature profile in the remediation zone is characterized by large gradients, and temperatures decrease sharply with distance from the source. Over time, superposition of heat from adjacent heaters tends to even out these differences.

Figure 2
Blanket and Thermal Well Heating

Blanket Heating

Thermal Well Heating

Source: TerraTherm™ Inc.

Vapor Extraction

As the matrix is heated, adsorbed and liquid-phase contaminants begin to vaporize. A significant portion of organic contaminants either oxidize (if sufficient air is present) or pyrolize once high soil temperatures are achieved. Desorbed contaminants are recovered through a network of vapor-extraction wells.

Vapor extraction wells are also heated to prevent condensation of contaminants inside the well. A vacuum is applied to these wells to induce air flow through the contaminated media creating a zone of capture. Contaminant vapors captured by the extraction wells are conveyed to an offgas treatment system for treatment prior to discharge to the atmosphere.

Offgas Treatment (Ref. 2)

TerraTherm offers two different methods of vapor treatment. One treats extracted vapor without phase separation (Figure 1), and the other cools heated vapor, separates the resulting phases, and manages each phase separately.

The vapor treatment option depicted by Figure 1 uses a thermal oxidizer to break down organic vapors to primarily carbon dioxide and water. Stack sampling has demonstrated that toxic pollutants in offgas, including dioxins, are substantially below regulatory standards. When influent vapors contain chlorinated compounds, hydrogen chloride (HCl) gas is produced. In such cases, the exhaust from the thermal oxidizer is passed through an acid gas scrubber to capture HCl gas.

The other vapor treatment option uses a heat exchanger to cool extracted vapors. The resulting liquid phase is then separated into aqueous and nonaqueous phases. The nonaqueous phase liquid (NAPL) is usually disposed of at a licensed treatment storage and disposal facility. The aqueous phase is passed through liquid-phase activated carbon adsorption units and then released into the environment. Cooled, uncondensed vapor is passed through vapor-phase activated carbon adsorption units and then vented to atmosphere.

Although setup varies from site to site, several components of the remediation system including heaters, blowers, and offgas treatment equipment are either standard or adaptable to new situations, with equipment reused from site to site. Downhole wells may not be salvageable and may be plugged and abandoned in place.

STATUS AND AVAILABILITY (REFS. 4 AND 5):
ISTD is a patented technology originally developed by Shell Oil. While U.S. Patent rights were donated to the University of Texas (UT), patent rights outside the U.S. were retained by Shell. TerraTherm holds the exclusive license to this technology from both UT and Shell, and is currently the only vendor. ISTD has been commercial for several years. Its ability to remove PCBs from contaminated soil was first demonstrated more than 6 years ago. As shown on Table 1, ISTD has been used at six POP-contaminated sites. Implementation at four of these sites was full scale, and the other two were pilot scale.

Table 1
Performance of ISTD at POPs Contaminated Sites (Refs. 2, 4 and 7)

Site Name	Year	Scale	Contaminant	Concentration			
				Initial	Final	Goal	Units
Former South Glens Falls Dragstrip, Moreau, New York	1996	Full	PCB 1248/1254	5,000 (Max)	< 0.8	2	mg/kg
Tanapag Village, Saipan, NMI	1997 - 1998	Full	PCB 1254/1260	10,000 (Max)	< 1	10	mg/kg
Centerville Beach, Ferndale, CA	1998 - 1999	Full	PCB 1254	860 (Max)	< 0.17	1	mg/kg
			Dioxins and Furans	3.2 (Max)	0.006 [1]	1	ug/kg TCDD
Missouri Electric Works, Cape Girardeau, MO	1997	Pilot	PCB 1260	20,000 (Max)	< 0.033	2	mg/kg
Former Mare Island Naval Shipyard, Vallejo, CA	1997	Pilot	PCB 1254/1260	2,200 (Max)	< 0.033	1	mg/kg
Former Wood Treatment Area, Alhambra, CA	2002 - 2005	Full	Dioxins	18 (Mean)	0.01	1	ug/kg

Note:

Avg   Average concentration
Max   Maximum concentration
mg/kg   Milligrams per kilogram (or parts per million)
NMI   Northern Mariana Islands
ND   Below detection limit
TCDD   Tetrachlorodibenzodioxin equivalents
ug/kg   Micrograms per kilogram (or parts per billion)

[1]   Final concentration presented as average of residual concentrations in treatment area.

DESIGN (REF. 12):
Key design factors for ISTD include the number and depth of heater wells and vacuum wells, as well as the requirements for electrical power and treatment of off gasses. These factors are affected by the type of contaminants present, concentration of the contaminants, extent of contamination, soil type, hydraulic conductivity, permeability, thermal properties, location of the water table, availability of site facilities such as electrical power supply, and regulatory issues.

THROUGHPUT (REF. 5):

ISTD has been used to treat volumes as low as a few hundred cubic yards to greater than 20,000 cubic yards in 6 to 9 months. Factors affecting cleanup durations can include type of contaminants, cleanup/remedial goals, and site geology.

WASTES/RESIDUALS (REFS. 3 AND 5):

Wastes produced by ISTD are likely to result from the treatment of extracted vapors, and vary according to the type of treatment they are subjected to. Offgas treatment options that employ phase separation techniques could produce process wastes such as NAPL, spent liquid- and vapor-phase activated carbon, and inorganic salts as waste products. For example, the treatment of chlorinated vapors in a thermal oxidizer results in the production of HCL gas. A wet or dry acid gas scrubber used to neutralize HCl gas will produce inorganic salts as a waste product.

NAPL is typically transported off site for disposal at a licensed facility. Spent activated carbon may either be disposed of, or regenerated at a licensed facility. Inorganic salts produced from neutralization processes are typically considered nonhazardous and are consequently disposed of as nonhazardous waste.

MAINTENANCE (REF. 4):

Maintenance associated with ISTD includes the occasional replacement of heater elements. ISTD operation is typically characterized by less than 5% downtime. Other maintenance needs include treatment media replacement and thermal oxidizer refueling.

LIMITATIONS (REF. 4):

The following are some of the limitations of this technology:

- ISTD cannot address contaminants that do not volatilize with in the temperature range of approximately 15-1000°C.
- As long as liquid water remains within the remediation zone, the temperature that can be attained is limited to the boiling point of water (212 °F). Once the water is boiled off, higher temperatures can be attained. A continuing source of water influx into the treatment zone will undermine the ability of this technology to produce temperatures necessary for the removal of POPs. For this reason, formation dewatering and implementation of water control measures are needed prior to the implementation of ISTD in high-permeability, water-saturated zones.
- Though not always the case, cost can be a limiting factor. Unit costs for treatment are influenced by several factors including scale of the project, depth of the treatment zone, depth to water table, air emission controls, cost of labor and cost of power. However, in general, unit costs in USD range from $200 to $600 per cubic yard corresponding to treatment volumes ranging from less than 5,000 to approximately 15,000 cubic yards for POP-type contaminants. Larger volumes may have lower unit costs. Treatment costs for VOC contaminants are lower.

FULL-SCALE TREATMENT EXAMPLES:

Centerville Beach (Refs. 6, 8, 10 and 14)

The Centerville Beach Naval Facility is a 30-acre site in Ferndale, California that was used for oceanographic research and undersea surveillance. The site was decommissioned in 1993. Operations at the site lead to contamination of a particular area with PCBs. The PCB of concern was Aroclor 1254 which was present in concentrations ranging from 0.15 to 860 milligrams per kilogram (mg/kg). Dioxins and furans were also present at a maximum concentration of 3.2 micrograms per kilogram (μg/kg) as 2,3,7,8-tetrachlorodibenzodioxin (TCDD) equivalents. The contaminated medium was primarily silty clay. Groundwater was encountered below the contaminated zone at depths exceeding 60 feet bgs.

From September 1998 through February 1999, approximately 1,000 cubic yards of PCB-contaminated soil was treated using ISTD. Heater and vapor extraction wells were installed in a zone measuring 40 feet long, 30 feet wide, and 15 feet deep. The wells were installed 6 feet apart. Two sealed vacuum blowers were used in parallel for vapor extraction. Offgas was treated using a flameless thermal oxidizer (with greater than 99.99% demonstrated treatment efficiency), and two granular activated carbon units configured in series. The total cost of the implementation in USD was approximately $650,000.

The treatment goal was 1 mg/kg for PCBs and 1 µg/kg TCDD equivalent for dioxins and furans. Remediation took place between September 1998 and February 1999. Treatment goals were met in the bulk of the treatment area; however, one portion (178 cubic yards) still contained elevated concentrations of PCBs. This was found to be caused by a previously undiscovered bank of PCB-containing electrical conduits emanating from outside the treatment zone and passed into the treatment area. Excavation and disposal was subsequently used to remove this area of contaminated soil and the associated conduits.

Alhambra (Refs. 3, 9, 17 and 18)

Southern California Edison's (SCE) Alhambra Combined Facility occupies approximately 33 acres and is currently used for storage, maintenance, and employee training. SCE carried out wood treatment operations in SCE's 2-acre former wood treatment area between 1921 and 1957. The total volume of contaminated soil was estimated to be 16,200 cubic yards of soil. The contaminated zone included a variety of buried features including treatment tanks, the structural remains of the former boiler house and tank farm, and various buried utilities. The contaminants of concern were PAHs, pentachlorophenol (PCP), and dioxins. Total PAHs were present in site soils at a maximum concentration of 35,000 mg/kg and an average concentration of 2,306 mg/kg. PCP was present at a maximum concentration of 58 mg/kg and an average concentration of less than 1 mg/kg. Dioxins were present at a maximum concentration of 0.194 mg/kg and an average concentration of 0.018 mg/kg (expressed as 2,3,7,8-tetrachloro-dibenzodioxin [TCDD] Toxic Equivalency Quotient [TEQ]). The soil in the remediation zone was composed of silty sands, inter-bedded with sands, silts, and clays. The average thermal treatment depth was approximately 20 feet bgs and extended to 100 feet bgs in some areas. The depth to the water table was greater than 240 feet bgs. The treatment goals were 0.065 mg/kg (expressed as benzo(a)pyrene [B(a)P] toxic equivalents) for PAHs; 2.5 mg/kg for PCP, and 0.001 mg/kg for dioxins (expressed as TEQ).

Remedial action at the site was conducted in two phases. Each phase addressed a different area of the site. The overall ISTD system for the two phases consisted of 785 thermal wells (131 heater-vacuum and 654 heater-only wells) at a 7.0-ft spacing between thermal wells, as well as an insulated surface seal, thermal oxidizer, heat exchanger, and granular activated carbon for off-gas treatment.

The ISTD began cleanup operations for Phase I of the remediation of Area of Concern (AOC)-2 in February 2003.

Confirmation soil samples were submitted to DTSC in July 2004 which confirmed that the cleanup goals for Phase I of AOC-2 had been achieved. Phase 2 of the cleanup began in July 2004 and was scheduled for completion by October 2004.

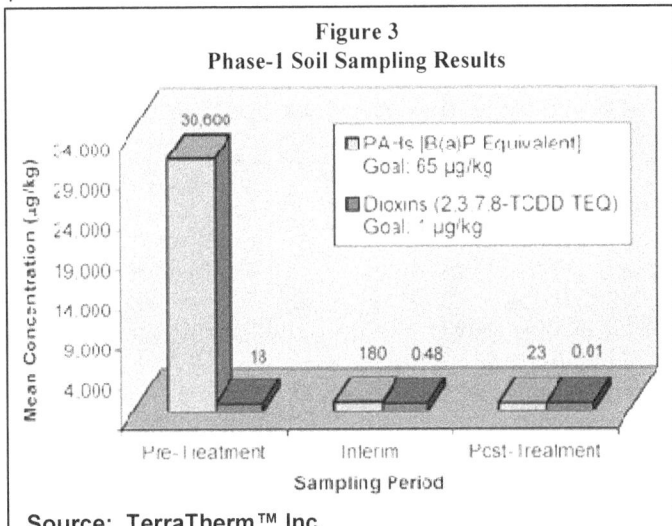

Figure 3
Phase-1 Soil Sampling Results

Source: TerraTherm™ Inc.

However, a previously undiscovered volume of free product made it necessary to reduce in-situ temperatures in order to control organic contaminant concentrations in the offgas treatment system influent. This resulted in an anticipated 10-month increase in the cleanup duration. Phase 2 of the cleanup is expected to end in August 2005. The total cost of the implementation in USD was approximately $10 million.

Rocky Mountain Arsenal Hex Pit (Ref. 15)

The Hex Pit was a former disposal pit at the U.S. Department of Army's Rocky Mountain Arsenal (RMA). Shell Oil Company leased a portion of the RMA from 1952 to 1982 to manufacture pesticides. The pit was used from 1947 to 1975 to dispose of residues from distillation and other processes used in the production of hexachlorocyclopentadiene (hex), an ingredient in the manufacture of pesticides.

The main part of the Hex Pit measured approximately 94 ft by 45 ft, and varied from 8 to 10 ft deep. The pit contained a total of 2,005 cubic yards of waste-contaminated materials, of which 833 cubic yards was estimated to be waste.

The Hex Pit consisted primarily of soil and waste material originally disposed of in the pit. The impacted soil (silty sand) was stained dark brown, rust orange, or black, and at times included granules or globules of hex. Black, tar-like, relatively pure hex residue occurred in distinct solid layers of waste (approximately 1-foot thick). Hex was not detected in groundwater downgradient of the Hex Pit boundaries.

The contaminants of concern were hex, aldrin, chlordane, dieldrin, endrin, and isodrin. Only hex, chlordane, and dieldrin had treatment goals. The treatment goals were 760 mg/kg, 67 mg/kg and 335 mg/kg respectively. Laboratory tests indicated that Hex Pit wastes could be effectively treated by the ISTD process.

ISTD at the Hex Pit was designed to heat a treatment soil volume of 3,198 cubic yards, extending from 0 to 12 ft bgs and 5 ft laterally beyond the boundaries of the Hex Pit. Thermal wells on 6-foot centers were installed in a hexagonal arrangement. A total of 266 wells were installed, of which 210 were heater-only and 56 were heater-vacuum wells.

The target treatment temperature based on the boiling point of COCs was 325 °C. All heater-only wells reached their operating temperatures in early March 2002. Treatment was expected to last 85 days and end in May 2002. However, twelve days after commencement, corrosion was observed in some of the well manifolds. Subsequent investigation and assessment determined that unforeseen concentration of HCL gas and production of HCL (liquid) in the vapor conveyance system, resulting from the highly concentrated wastes in the Hex Pit, had caused corrosion. Corrosion damage to the ISTD system was significant. A determination was made that replacements with necessary corrosion resisting matrices was cost prohibitive. Wastes were excavated and capped.

STATE CONTACT (CENTERVILLE BEACH):	STATE CONTACT (ALHAMBRA):	VENDOR CONTACT:
California EPA Dept. of Toxic Substances Control (DTSC) Ms. Christine Parent Phone: (916) 255-3707 Email: CParent@dtsc.ca.gov	California EPA DTSC Mr. Tedd E. Yargeau Phone: (818) 551-2864 Email: tyargeau@dtsc.ca.gov	Mr. Ralph Baker TerraTherm™, Inc. Tel: (978) 343-0300 Email: rbaker@terratherm.com

PATENT NOTICE:

ISTD is covered by a total of 22 U.S. patents, with 6 patents pending. TerraTherm is the exclusive licensee through the University of Texas and Shell.

REFERENCES:

1. Baker, Ralph and Kuhlman, Myron. 2002. 2nd International Conf. on Oxidation and Reduction Technologies for Soil and Groundwater, ORTs-2, Toronto, Ontario, Canada. A Description of the Mechanisms of In-Situ Thermal Destruction (ISTD) Reactions. Nov. 17-21

2. Baker, Ralph, TerraTherm, Inc. 2004. Email to Chitranjan Christian, Tetra Tech EM Inc., Regarding Questions on ISTD. October 27, November 8, 15, 24 and 29.

3. Baker, Ralph, TerraTherm, Inc. 2004. Telephone Conversation with Chitranjan Christian, Tetra Tech EM Inc., Regarding Questions on ISTD. October 29.

4. Heron, Gorm, TerraTherm, Inc. 2004. Email to Chitranjan Christian, Tetra Tech EM Inc., Regarding Questions on ISTD. October 15.

5. Heron, Gorm, TerraTherm, Inc. 2004. Telephone Conversation with Chitranjan Christian, Tetra Tech EM Inc., Regarding Questions on ISTD. October 15.

6. Parent, Christine, California EPA, DTSC. 2004. Telephone Conversation with Chitranjan Christian, Tetra Tech EM Inc., Regarding Questions on ISTD implementation at Centerville Beach. November 2.

7. Stegemeier, G.L., and Vinegar, H.J. 2001. "Thermal Conduction Heating for In-Situ Thermal Desorption of Soils," Chapter 4.6, pages 1-37. Chang H. Oh (ed.), Hazardous and Radioactive Waste Treatment Technologies Handbook, CRC Press, Boca Raton, FL.

8. TerraTherm Environmental Services. 1999. Naval Facility Centerville Beach, Technology Demonstration Report, In-Situ Thermal Desorption. November.

9. TerraTherm Inc. Case Study – Alhambra. Online Address: http://www.terratherm.com/CaseStudies/WS%20Final%20Alhambra%20Sheet.pdf.

10. TerraTherm Inc. Case Study – Centerville Beach Naval Facility. Online Address: http://www.terratherm.com/CaseStudies/WS%20Centrvll-Tesi.pdf.

11. TerraTherm Inc. Case Study – Former Mare Island Naval Shipyard. Online Address: http://www.terratherm.com/CaseStudies/WS%20BADCAT.pdf.

12. TerraTherm Inc. Feasibility Screening. Online Address: http://www.terratherm.com/default.htm.

13. TerraTherm Inc. ISTD Process Description. Online Address: http://www.terratherm.com/default.htm.

14. Tetra Tech EM Inc. 2000. Draft Final Closeout Report. Naval Facility Centerville Beach, Ferndale, California. February.

15. Todd, Levi. Year. Publication or Report. Lessons Learned from the Application of In Situ Thermal Destruction of Hexachlorocyclopentadiene Waste at the Rocky Mountain Arsenal. Month.

16. U.S. Environmental Protection Agency. 2004. In Situ Thermal Treatment of Chlorinated Solvents Fundamentals and Field Applications. EPA 542-R-04-010. March.

17. Yargeau, Tedd, California EPA, DTSC. 2004. Email to Chitranjan Christian, Tetra Tech EM Inc., Regarding Questions on ISTD implementation at Alhambra. December 22.

18. Yargeau, Tedd, California EPA, DTSC. 2004. Telephone Conversation with Chitranjan Christian, Tetra Tech EM Inc. Response to Questions on ISTD implementation at Alhambra. November 2.